Stainless Steel Surfaces

ZAHNER'S ARCHITECTURAL METALS SERIES

Zahner's Architectural Metals Series offers in-depth coverage of metals used in architecture and art today. Metals in architecture are selected for their durability, strength, and resistance to weather. The metals covered in this series are used extensively in the built environments that make up our world and are also finding appeal and fascination to the artist. These heavily illustrated guides offer comprehensive coverage of how each metal is used in creating surfaces for building exteriors, interiors, and art sculpture. This series provides architects, metal fabricators and developers, design professionals, and students in architecture and design programs with a logical framework for the selection and use of metallic building materials. Forthcoming books in *Zahner's Architectural Metals Series* will include Stainless Steel; Aluminum; Copper, Brass, and Bronze; Steel; and Zinc surfaces.

Titles in *Zahner's Architectural Metals Series* include:

Stainless Steel Surfaces: A Guide to Alloys, Finishes, Fabrication and Maintenance in Architecture and Art

Stainless Steel Surfaces

A Guide to Alloys, Finishes, Fabrication, and Maintenance in Architecture and Art

L. William Zahner

WILEY

Cover Design: Wiley
Cover Image: © releon8211/iStockphoto
Designed by Kohn Pedersen Fox Architects.
Photo by Tex Jernigan | ARKO.

This book is printed on acid-free paper.

Published by John Wiley & Sons, Inc., Hoboken, New Jersey.

Published simultaneously in Canada.

ISBN 9781119541547 (paper); ISBN 9781119541608 (ePDF); ISBN 978111951585 (ePub)

Printed in the United States of America

V10012144_071519

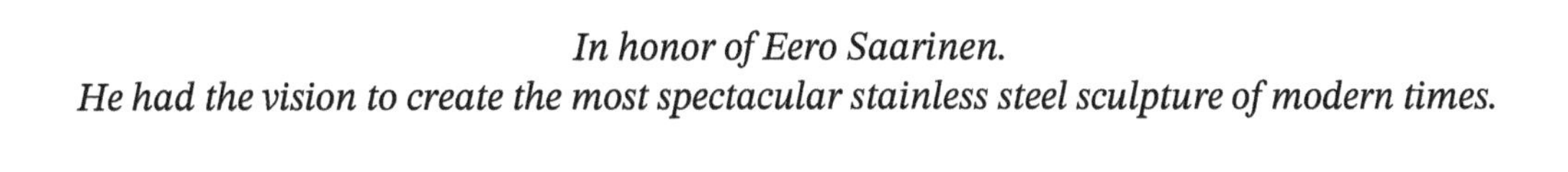

In honor of Eero Saarinen.
He had the vision to create the most spectacular stainless steel sculpture of modern times.

Contents

Preface

Felix qui potuit rerum cognoscere causus (Fortunate is he that knows the causes of things).

Virgil

When I began my career working in the architectural metal industry, I realized my knowledge was deficient when it came to metals in architecture. What I found, however, was that the knowledge in the design community was also paltry. There were a lot of colloquial "guidelines" based on best practice or "rule of thumb" generalizations. Corrosion information was like a feudal cast system: one metal was deemed more noble than the next based on some obtuse galvanic series lists of metals *in flowing sea water*. Unless you were designing oceangoing vessels, what does that have to do with architecture? I can recall working at a power plant where my firm was installing painted steel siding. The inspector walked by and dropped a magnet in the bucket of stainless steel fasteners and saw that they were slightly magnetic. From this, the inspector concluded that we were using the wrong fastener. "They must be steel," he said, and he had the work halted until I could prove to him that even austenitic stainless steels could be magnetic. He was the "expert," but his knowledge of metals was limited. This can be costly.

The finishes on stainless steel that were available and being promoted, displayed in little packages of neat samples, were limited to mill finishes and finishes induced by parallel scratches or polished to a poor mirror. For sheet forms of the metal there were but six or seven finishes in regular use at the time in North America. These were the finishes to choose from. Industry said so.

Coloring of stainless steels did not arrive commercially until the late 1970s and this was mainly for signage. The Japanese embraced coloring of stainless steel in the art and architecture market, but it really did not expand commercially as an architectural surface before Antoine Predock produced his design for the Museum of Science and Industry in Tampa, Florida, and Frank Gehry designed the Euro Disney Resort in Paris and the Team Disney Anaheim building in Anaheim, California. The first major use of glass bead finish stainless steel in the United States was the IBM headquarters design by Kohn Pedersen Fox. Gehry's Team Disney Anaheim building used glass bead stainless steel with a green interference color. The first major sculpture to use colored stainless steel was *Ad Astra* by Richard Lippold, which was installed in front of the Smithsonian's National Air and Space Museum. These and others sparked the beginning of development and helped pave the way for making stainless steel a major material of design.

When I first considered writing a book on the metals used in architecture it was from a burning desire to learn more about this particular form of matter. I wanted to gain an understanding of the constraints as well as what supported the available information on metals. Architectural metal had been the business of the three previous generations of my family. Much of what we created over those generations was confined to the knowledge of the time, much of it limited the use of metals on thin roof surfaces and painted roll formed panels. I wanted to understand the use of metals better, and, perhaps, part of me wanted to explore and test and push the knowledge to new areas of understanding. Further research for our clients and for myself led me to write my first book, *Architectural Metals: A Guide to Selection, Specification, and Performance*, back in the early 1990s.

Since writing that book, I have had the honor to work with many incredible designers in achieving phenomenal works in metal – works that will stand the test of time and provide inspiration to future generations. I have had the opportunity to investigate surface conditions on many works of art and determine the means of restoration, extending their useful life as a form of art and beauty. Experience, research, and discourse with the design and engineering community has broadened the conversation and with it the boundaries of knowledge.

New techniques in forming and finishing metals coupled with the sheer effort of creating by the design and art community has crushed the normative boundaries that limited our experiences. We are in a new age: the *Ad Astra* movement. Conventions and generally accepted opinions that claim to establish the bulk of knowledge about a material or process are being overwhelmed by new evidence – by physical proof, a proof that is often remarkably beautiful.

A favorite koan from the Zen master Shunryu Suzuki says, "In the beginner's mind there are many possibilities, but in the expert's there are few." It is the artist in us that challenges the conventions and pushes the edges. In this way we are the beginner. The artist is not confined by the constraints of manufacturing norms. Like the scientist, he or she seeks new boundaries defined by knowledge.

When I considered an update to the book, I soon found that the information on stainless steel alone would today easily fill its own volume. Stainless steel may be relatively new as metals go, but the plethora of finishes and colors that can be applied to the metal give stainless steel a versatility few materials can match. The first book of this series on metals is on the king of the architectural metals, stainless steel. It is the king because it has so many of the attributes that artists and designers seek in a material. (In a chessboard sense it should be the queen, the most versatile and powerful piece on the board).

My hope is that some of the knowledge I have gained on the metal will inspire designers and artists to consider the steel we call stainless steel and work to push constraints out a bit further. We have not found the edge yet.

L. William Zahner

CHAPTER 1

Introduction to Stainless Steel

All that glitters is not gold.

William Shakespeare, *The Merchant of Venice*

STAINLESS STEEL

Few metals can compete with the versatility of stainless steel. Stainless steel, in the hands of a knowledgeable fabricator, offers infinite variations of finishes and textures that can be mixed, layered, etched, and embossed to create a surface appearance and reflectivity no other materials can match. Add to this the strength, hardness, and corrosion resistance stainless steel offers and you have a very special material of design.

Stainless steel is relatively new as far as being used in art and architecture. It does not have the long history of copper and bronze. It was not known at the time of Napoleon, when zinc became the metal of Paris architecture. Even aluminum was in use before stainless steel came into prominence. Its discovery and subsequent utility was not known prior to the twentieth century, and its humble start was its use in special nonrusting cutlery.

Today, stainless steel is synonymous with the modern age. Art conservationists consider it one of the so-called modern metals. Stainless steel's jewel-like luster adorns building facades and roofs on many major structures around the world. As a sculptural material, stainless steel is accepted and used by some of the most talented artists of our time.

> Few metals can compete with the versatility of stainless steel.
>
> Stainless steel, in the hands of a knowledgeable fabricator, offers infinite variations of finishes and textures that can be mixed, layered, etched, and embossed to create a surface appearance and reflectivity no other material can match.

Perhaps the most compelling reason to consider using stainless steel to visually express an art form or adorn a building surface is the luster of the metal. The luster of stainless steel is rich and intense. Unlike the warmth attributed to silver and pewter, stainless steel conveys a sense of modern achievement as a material that can stand the test of time.

The color is attributed to the chromium in the alloy. Chromium gives stainless steel a bluish tone. Some would consider this tone cold and sterile, but when stainless steel is given a finishing texture of minute scratches or tiny indented glass bead–blasted craters, the metal captures the light and sends it back to us in ways that no other material can.

Stainless steel is one of the few metals that is expected to stand against nature and not change in perceivable ways. Stainless steel is homogeneous in its makeup. Scratch it, cut it, or pierce it, and you have the same stainless steel throughout. It does not require other metals to protect it, and stainless steel does not need paint to seal its surface from the rigors of the environment.

Stainless steel is a noble metal in its passive state. By passive state we mean a nonreacting and stable state. The passive state is one stainless steel resides in for most applications insofar as art and architecture are considered. Because of this passive state, bolts and fasteners made from stainless steel support many steel and aluminum structures. Even copper roofs can be held down with stainless steel clips without fear of galvanic reaction. Like titanium, another metal even more recently introduced to the cladding scene, thin skins of stainless steel can enclose structures and be expected to last hundreds of years with little maintenance to aid that which nature provides.

Stainless steel reflects around 60% of the visible light spectrum. It absorbs the shorter wavelengths of the visible spectrum, which is the blue end. One would think that if it absorbs more of the blue spectrum, then the color reflected would be reddish, but that is not so. Stainless steel has a markedly bluish tint. With metals, the greater the absorption, the greater the reflection. When light strikes a metal surface it is intensely absorbed to a level of a few hundred atoms. These atoms react to the light by exciting the electrons on the surface. Light is an electromagnetic wave, so the electrons in the surface atoms of the metal develop alternating currents, which release energy in the form of light and create a strong reflection of the absorbed wavelength. Stainless steel absorbs more of the blue wavelength and less of the red or yellow. Stainless steel absorbs light energy where it excites the electrons and reemits a strong metallic appearance that is more intensely blue.

Another valuable aspect of stainless steel is its incredible corrosion resistance. There are few other metals that show such a disregard for the ravages of the environment. Stainless steel is slow to change – unless confronted by chlorides. Chlorides leave their mark on stainless steel. Chlorides have a proclivity for metals and stainless steel is no exception. Chlorides can attack a metal surface and break down the bonds that bind them into solid forms. There are alloys, however, that have been developed to fend off the attack that can occur when stainless steels are exposed to chloride and other damaging environments.

This corrosion resistance characteristic of stainless steel is the attribute that enables the metal to be a changeless design material. Stainless steel surfaces remain unchanged in the typical environment of our cities and towns. This leads to a predictability and a consistency in the unique metallic appearance stainless steel possesses.

Stainless steel used as a cladding material inside or outside can be enhanced further by the choice of the finishing texture. These finishes essentially paint with light. A skillful choice of finish can enhance what the eye will see as the full potential of the luster of stainless steel is brought out. Designers and artists have a multitude of choices and combinations when working with stainless steel to emphasize their creations (Figure 1.1).

The choice of stainless steel as a design component brings to bear the confluence of nature and technology. Stainless acts as a natural material. From the earth and the sky in the form of meteorites, you have iron and nickel, two common metals that add strength and resilience. These metals and their strength have been known to civilizations for ages. Through a century of discovery and analysis, man added chromium, another metal, along with a tremendous amount of manmade energy, to arrive at what we call stainless steel: an alloy of iron, chromium, and for architecture, nickel. Stainless steel is homogeneous throughout, but like the noble metals copper, gold, silver, and platinum found naturally in a pure state, this manmade metal has a thin coating of clear oxide that gives it the ability to resist oxidation and change in a similar noble way. The thin oxide is a chromium-rich barrier that enables stainless steel to possess the strength and durability of steel and the ductility and toughness of nickel.

Modern forms of stainless steel utilize scores of different elements in different combinations to arrive at specialized, well-tuned characteristics. For art and architectural purposes, the range of alloys is not so vast. In Chapter 2, those alloys that play a role in art and architecture and a few that reside on the edges of potential consideration are described in more detail (Table 1.1).

The visual attributes are provided by the finish and texture of the surface and all of the alloys listed can be finished in similar ways. The alloy selection, for the most part, is to aid performance in the context of selected natural exposures in which corrosion inhibition needs a boost or fabrication techniques require subtle adjustments. It is important to note that the industry uses the term "grade" – as in different grades of stainless steel – to identify different alloy families. There are four main families of stainless steel: ferritic, austenitic, martensitic, and duplex. These are distinguished by the metallurgic structures of their crystal forms (Figure 1.2). All metals are made of crystals. Crystals form the matrix of atoms that define much of the mechanical behavior of the metal. For stainless steel, each structure contributes various mechanical attributes to the metal. For instance, the body-centered characteristic of ferritic stainless steels make them magnetic like steels, whereas the austenitic alloys of stainless steel are nonmagnetic.[1]

Stainless steel is the name given to a group of steel alloys that contain various levels of chromium. At least 10.5% chromium is needed for a steel alloy to be included in the family of stainless steels. Other elements, such as nickel and molybdenum, are added in various amounts to give this special steel alloy attributes of value for art and architecture.

[1]In an annealed condition, austenitic alloys are nonmagnetic. Cold working these alloys will develop magnetic properties, as their crystals become distorted. Chapter 2 goes into deeper discussion about the magnetic permeability of the various stainless steel alloys.

FIGURE 1.1 Roxy Paine's sculpture in front of the Eli and Edythe Broad Art Museum. Source: Photo courtesy of Justin Maconochie Photography.

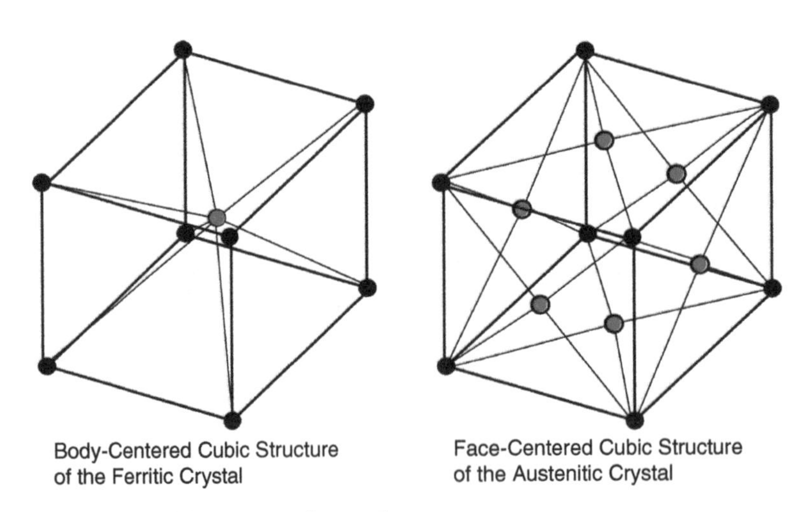

FIGURE 1.2 Body-centered and face-centered crystal structures.

TABLE 1.1 Selected alloys of stainless steel.

UNS No.	Metallurgic class	Common reference	European EN No.	Common use
Wrought				
S43000	Ferritic	430	1.4016	Food service
S43932	Ferritic	439LT	1.4509	Exterior, exhaust systems
S44400	Ferritic	444	1.4521	Food processing
S17400	Martensitic	17-4 PH	1.4542	Bolts
S41000	Martensitic	410	1.4006	Fasteners
S42000	Martensitic	420	1.4021	Cutting blades
S20100	Austenitic	201	1.4372	Springs, food service
S20200	Austenitic	202	1.4373	Exterior applications
S30300	Austenitic	303	1.4305	Machined
S30323	Austenitic	303Se	1.4325	Machined
S30400	**Austenitic**	**304**	**1.4301**	**Exterior and interior cladding**
S30403	**Austenitic**	**304L**	**1.4307**	**Exterior and interior cladding**
S31600	**Austenitic**	**316**	**1.4401**	**Exterior and interior cladding**
S31603	**Austenitic**	**316L**	**1.4404**	**Exterior and interior cladding**
S31703	Austenitic	317L	1.4438	Corrosive environments
S31254	Austenitic	312L	1.4547	Corrosive environments
S32001	Duplex	2001	1.4482	Corrosive environments
S31803	**Duplex**	**2205**	**1.4462**	**Chloride environments**
Cast				
J92500	**Austenitic**	**304L**		**Casting elements**
J92600	**Austenitic**	**304**		**Casting elements**
J92701	Austenitic	303Se		Machining
J92800	**Austenitic**	**316L**		**Casting elements**
J92900	**Austenitic**	**316**		**Casting elements**
J92205	Duplex	2205		Corrosive environments

Bolded alloys are common art and architecture alloys.

This chromium-alloyed steel, given the name "stainless," possesses the property of a very stable, unchanging surface. Unfortunately, to the uninitiated the name is sometimes misleading. Stainless steel gets its "stainless" character from the thin chromium oxide film that forms over the entire surface when it is exposed to the atmosphere. The face side, reverse side, and edges all develop this thin, unbroken chromium oxide protective layer. If, however, this film is damaged or breached by exposure to certain corrosive environments, stainless steel will perform no better than a high grade of steel. When exposure to deicing salts and humid coastal conditions change the surface appearance of stainless steel by its forming first a thin dark stain and then spots of dark red, the first thought is that the metal is not stainless steel but some inferior substitute. Once the stains are cleaned and the salts removed, the stainless steel will return to its original luster. There is no chicanery in the naming of this metal. It truly is a remarkable metal of our modern age. One only needs to perform an occasion cleaning and prevent mainly chloride ions from remaining on the surface for a long period of time (Figure 1.3).

There are many surface treatments that have been developed for stainless steel. These treatments, both mechanical and chemical, take advantage of the reflectivity afforded by the chromium

FIGURE 1.3 The Chrysler Building, designed by William Van Alen.

TABLE 1.2 Surface finishes.

Category	Description	Applicable form
Mill	2D	Sheet, plate, rod, tube, pipe
	2B	Sheet, plate, rod, tube, pipe
	2BA	Sheet
Mechanical-directional	3	Sheet, plate, rod, tube, pipe
	4	Sheet, plate, rod, tube, pipe
	Hairline	Sheet, plate, rod, tube, pipe
Mechanical-non-directional	Angel Hair	Sheet, plate, rod, tube, pipe, cast
	Glass bead	Sheet, plate, rod, tube, pipe, cast
	Shot blast	Sheet, plate, rod, tube, pipe, cast
	Swirl	Sheet, plate, rod, tube, pipe, cast
	CrossFire	Sheet
Mechanical-mirror	7	Sheet, plate, rod, tube, pipe
	8	Sheet, plate, rod, tube, pipe
	Super Mirror	Sheet
Plastic deformation	Water	Sheet
	Embossed	Sheet
	Selective embossed	Sheet, plate
	Coining	Sheet
Chemical	Thermal	Sheet, plate, rod, tube, pipe, cast
	Interference	Sheet, plate, rod, tube, pipe, cast
	Physical vapor deposition	Sheet
	Electropolish	Sheet, plate, rod, tube, pipe, cast
	Acid etch	Sheet

in the alloy and by the consistency and predictability of the surface (Table 1.2). The jewel-like metallic luster is one of the principle characteristics of the metal. Because the surface is relatively inert, the luster does not change with the development of the chromium oxide layer. You can mechanically alter the surface reflection in a number of ways without damaging the inert character of the metal. This is the main attribute that makes stainless steel so desirable. Innumerable finishes and textures can be introduced to the surface of stainless steel in order to take advantage of this beautiful, consistent luster (Figure 1.4).

FIGURE 1.4 *Hope for Life®* sculpture, by Larry Young.

Stainless steel can be given color from controlled development of interference oxides or from physical vapor deposition (PVD) coatings that also produce a thin interference oxide on the surface. The colors are limited by the reflective index of the film or oxide, but they can be enhanced further by first applying various mechanical surface finishes to the base metal surface (Figure 1.5). This capability makes stainless steel one of the more versatile metals for art and architecture design. The metal surface's ability to resist change when exposed to ambient conditions allows for a predictable appearance that is sustainable for a long period of time. No further barrier coatings are necessary, and the mineral nature of the surface does not change as a result of exposure to ultraviolet radiation from sunlight.

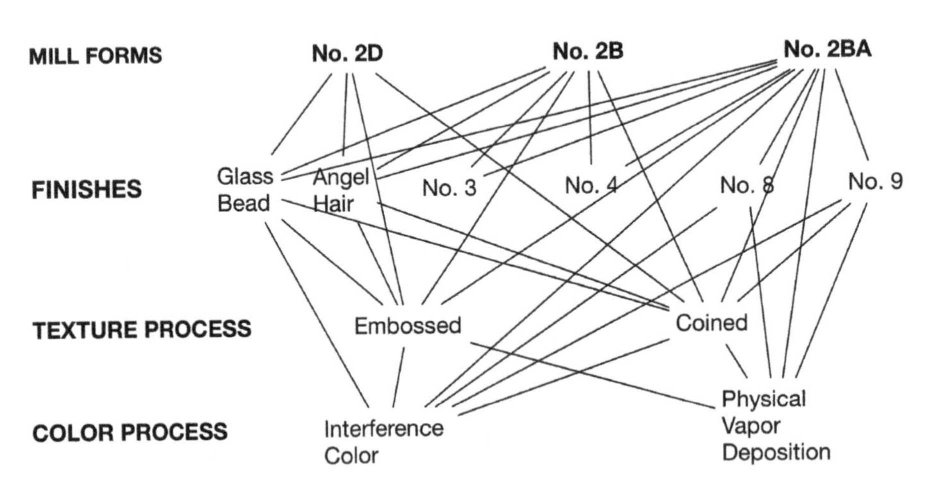

FIGURE 1.5 Combinations of finishes that can be applied to wrought stainless steel.

HISTORY

Stainless steel came into use as an architectural surfacing material in late 1920s, the first major work being the Chrysler Building in New York. The Chrysler Building is clad in stainless steel with an alloy content resembling alloy S30200. Prior to that, art deco surfaces with a silvery look were composed of nickel-bearing alloys of copper, such as nickel silver, German silver, and Monel. To a lesser degree, aluminum was used for interior art deco ornamentation. Aluminum was discovered only a few decades prior to the discovery and commercial use of stainless steel. These special alloy steels, iron alloyed with chromium, were found to retain a shiny appearance and resisted tarnishing in normal atmospheres. Further testing showed they possessed a resistance to rusting unlike other steel alloys. These special steels resisted oxidation, yet still possessed exceptional toughness.

For the most part, these first developments involved several different people in different parts of the world and happened in the early years of the twentieth century. The person who actually "invented" stainless steel is up for debate. Some credit Harry Brearley of England. In 1913 Brearley created a steel composed of 12.8% chromium. This chromium steel was used early on for cutlery, because it could hold a sharp edge without corroding. Brearley called it "rustless steel." In a trade journal, it was later described as stainless steel by Ernest Stuart of the R.E. Mosley cutlery company. The name stuck and was used to describe this new steel that when forged into cutlery would not corrode.

Some say that the invention of the steel alloy took place much earlier and that it was developed by the brilliant scientist Michael Faraday. He and James Stodart worked on various iron and alloying elements as far back as 1821. Their experiments involved the addition of various metals, such as platinum and nickel, as well as chromium. Their research was intended to find better cutting steels and to ascertain whether a steel could be created that was less susceptible to corrosion. Faraday and Stodart added as much as 3% chromium to the steel in their experiments. In 1904, the French metallurgist Leon Guillet developed iron-chromium alloys and iron-chromium-nickel alloys that were very close to the ferritic and austenitic alloys we still use today. Then there is the German physicist Dr. Benno Strauss, who worked with Dr. Eduard Maurer. Together they applied for a patent on a rust-resistant steel called V2A, which was a product of Krupp Steel Works. This later was called Nirosta, from the German term nichtrostender Stahl, which means "nonrusting steel." Nirosta was the type of steel used on the Chrysler Building.

In the United States around this same time, there were several people involved with the development of steel alloys high in chromium. Elwood Haynes started the Haynes Stellite Works in 1914 and produced a chromium-tungsten steel alloy. In 1907 he filed for and was awarded a US patent, No. 873,745, on chromium and cobalt alloy steels. In 1919, Haynes was awarded a patent for a specific range of chromium steels when he submitted several samples of his stainless steel to the patent office. He eventually sold his US patents to Harry Brearley, who started the American Stainless Steel Company.

This new metal – that had the strength of steel but was corrosion resistant, possessed a luster that rivaled silver yet would not tarnish, and that was heat resistant, hard yet malleable, and affordable – entered the realm of industry. Stainless steel moved mankind to another evolutionary stage of industrial revolution. These early pioneers saw the importance of such a steel. These advances

Rustless Steel
Cutlery Steel
Stainless Iron
Ascoloy
Firth's Aeroplane Steel
Jessop Hi-Gloss
Circle L No. 19
Anka
"It Stains Less"
Staybrite
Vesuvius
No-Kor-O-H
Atlas Stainless
Enduro S
No. 17 Metal

Various names for stainless steel in the early years of development

FIGURE 1.6 Early trade names for stainless steel.

in metallurgy mark a period of time in the mastery of metals by mankind that has never before or since been equaled.

The uses for stainless steel exploded after the early 1920s, as the number of manufacturers and the advancement of metallurgical knowledge extended the beneficial characteristics of these chromium-bearing steels. The early stainless steel alloys that found use in cutlery by Harry Brearley lacked ductility and could not be welded effectively. It wasn't until the alloys were refined in Sheffield, England, using the metallurgical work of Dr. Strauss in Germany, that more useful and malleable forms of stainless were created for commercial use.

As mentioned previously, early chromium-steel alloys had numerous trade names, such as V2A and Nirosta from the Krupp Steel Works, before eventually the name stainless steel was settled on. In these early years, the United States alone had nearly 100 companies making different forms of stainless steel. Each had its own trade name associated with the various forms of the metal produced. Some of the early trade names used are listed in Figure 1.6.

Today, the name stainless steel is ubiquitous across most of the world as a term for chromium-bearing steels. In some regions of Europe, stainless steel is known instead as "Inox" (after the French term "inoxydable") or as "Rostfrei" (German for "rust free").

MODERN PRODUCTION

Uses of the metal began with the novel and useful cutlery of Harry Brearley and the Firth Company's stainless knives, but quickly moved into automobiles, turbines, aircraft, and oil exploration equipment, where toughness and corrosion resistance are paramount. Architecture and art were not far

FIGURE 1.7 The St. Louis Arch, designed by Eero Saarinen.

behind. From the William Van Alen's art deco spire on the Chrysler Building to Eero Saarinen's 192-meter-tall Gateway Arch, stainless steel was the design material that would be synonymous with the modern age of mankind. Humans could proclaim that they had conquered nature by creating a substance that would not yield to the relentless onslaught of weather and time (Figure 1.7).

St. Louis's Gateway Arch was completed in 1965. The metal was supplied by two stainless steel mills, US Steel's Homestead Steel Works in Pittsburgh, Pennsylvania, and Eastern Stainless Steel's plant in Baltimore, Maryland. The arch is made of 886 tn. of S30400 stainless steel and 6 mm (1/4 in.) thick plates with a directional finish of approximately 180 grit. At the time the stainless steel was produced for the arch, standard mill practice was to use a basic electric arc furnace to pour the castings needed to be rolled into the large plates that make up the beautiful skin of the arch. All stainless steel used in architecture prior to 1968 was produced using the electric arc furnace method. The stainless steel used on the arch acts as a structural skin. There is an interior skin of steel plates, the lower 70 m of which have a concrete core. The thick stainless steel skin is fully welded with stiffeners and plug welded on the interior, concealed surface.

In 1968, the Union Carbide Corporation commercialized a method of producing stainless steel with superior metallurgical attributes. The method is known as the argon-oxygen decarburization process (commonly called the AOD process). Union Carbide developed this technique to improve the quality of stainless steel in the 1950s, but it was not put into use for another decade. This process and a similar process in Japan, known as vacuum-oxygen decarburization, are used to produce the stainless steel we find in art and architecture today.

The benefit this process brought to the production of stainless steel was a significant reduction in carbon and tighter control of the chemistry of stainless steel. The electric arc furnace method could produce a stainless steel with carbon as low as 0.03%. This new method produced stainless steels with carbon as low as 0.01–0.02%. A further benefit was the removal of sulfur. Using the AOD process, mills could provide a stainless steel with sulfur as low as 0.0005%. The reduction of carbon allowed for cleaner welding conditions and improved corrosion resistance at the welds. The reduction of sulfur improved the corrosion resistance and improved the clarity of the stainless steel surface.

The AOD process involves taking the melted scrap and virgin alloying material and placing them in a vessel that allows oxygen and nitrogen, then oxygen and argon, to be pumped through the melted metal. The ratio of oxygen to the other elements is stepped down as the carbon is removed. Once the carbon is removed and the desired nitrogen levels are achieved, other elements – aluminum, silicon, and calcium – are added to reduce the sulfur (Figure 1.8).

These processes greatly enhance the quality and corrosion resistance of this type of steel, in particular when welding is performed. The low carbon content reduces the chromium carbide behavior that occurs when the heat of welding draws the carbon out of the solid solution of the steel.

Reduction of sulfur is a significant benefit to artistic and architectural uses of the metal. As stainless steel is polished, the sulfur intermetallic will be exposed on the surface. This will interfere

FIGURE 1.8 An AOD chamber.

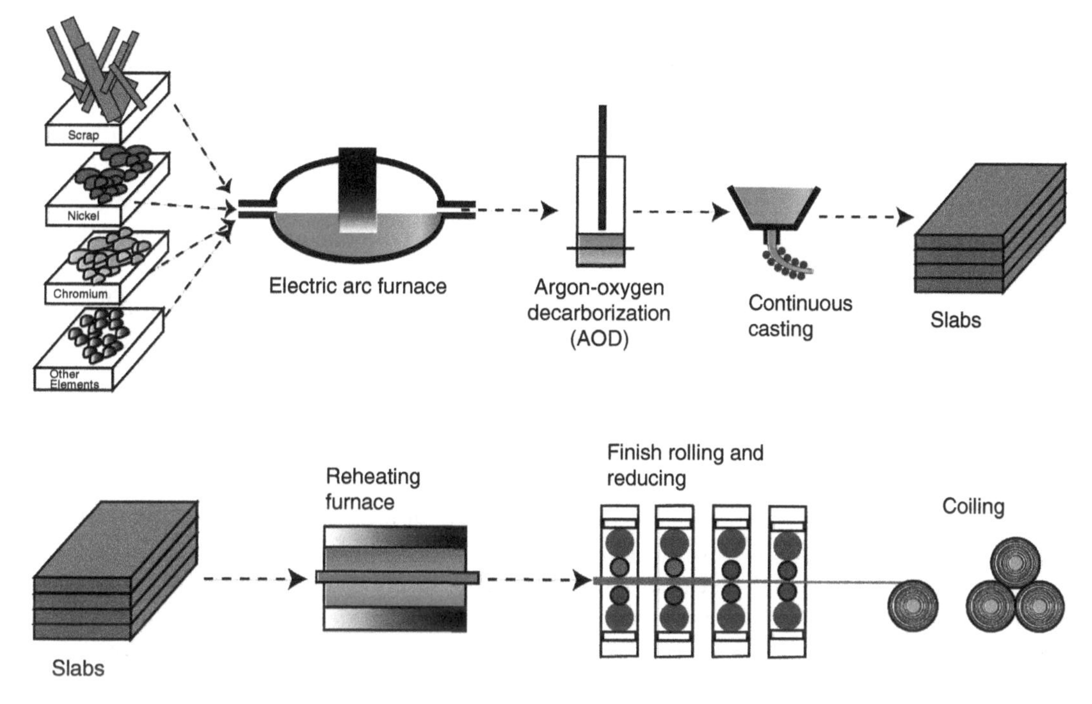

FIGURE 1.9 Diagram of mill production process.

with the clarity of the reflective luster. Removing sulfur enables the surface of stainless steel to receive a superior finish appearance. The sulfur can also lead to small surface pitting, so its removal improves the corrosion resistance character of the metal (Figure 1.9).

ENVIRONMENTAL, SAFETY, AND HYGIENE

Stainless steel alloys have no toxicity in the forms used in art and architecture. Alloys behave differently from their constituent elements. The chromium and nickel, as well as other substances, are tightly bonded in the metal matrix that forms the alloy. Under normal and expected exposures and use of stainless steel, no release of chromium or other elements will occur.

Stainless steel alloys are considered hygienic and are acceptable for uses related to drinking water, food preparation, dairy operations, and medical devices, including implants. Stainless steel is used in hospitals, in countertops in restaurants, in bathroom fixtures, and in sinks and clean rooms. Stainless steel is easy to clean and maintain. Its oxide is clear and adherent. Stainless steel does not promote bacterial growth, nor does it provide a surface for bacteria to collect and flourish. Stainless steel is insoluble in water and the passive oxide layer resists dissolution in most environmental conditions posed by or experienced by human interaction.

Chromium and nickel are the metal elements that, if not alloyed, can pose a hazard, but in the alloy form of stainless steels these elements are bound. Even with powder forms of stainless steel the elemental makeup does not pose a hazard, because of this metallic fixation that alloying creates.

Fumes from welding operations and dust from grinding operations should be avoided. Use of sulfur-bearing alloys has shown that there could be an increase in the release of nickel close to maximums allowed by European environmental codes. The sulfur-bearing alloys are used when significant machining is needed. The advent of the AOD process has greatly reduced sulfur in the alloys typically considered for art and architectural fabrication.

Stainless steel is recyclable. Recycling of stainless steel has grown dramatically and has become the main raw source of the metal. The International Stainless Steel Forum states that objects made of stainless steel today have an average recycled content of approximately 60%. Scrap recycling uses less energy. The recycled scrap is melted in electric arc furnaces. The electricity needed to melt the scrap does produce carbon dioxide emissions at the power generation plant.

Stainless steel is recycled on an ongoing basis. The value of stainless steel scrap is higher than steel, thus very little makes it to the trash heap. In addition, objects and surfaces created from stainless steel have a long projected life cycle. Once an object's useful life is over, the stainless steel in it will most certainly be recycled. Because stainless steel is rarely coated with paint, it can be readily recycled and put back to use. Interference colored stainless steels have no added substances that would require removal. The color is from light-wave interference, not dyes or pigments. These can be recycled as typical stainless steels. The combination of the expected long useful life of the metal coupled with its recyclability make stainless steel exceptional among materials used in art and architecture.

From a hygienic standpoint, most stainless steel surfaces are as easy to clean as glass or ceramic surfaces. This is because the surface does not develop a roughened oxide, nor is it very porous. Removal of soils and bacteria using common cleaners is possible with stainless steel surfaces. Seams where substances can be trapped can be eliminated or designed around by considering a welded and polished joint, making a monolithic, seamless surface. Tests on stainless steel surfaces have shown that the washing time versus the number of bacteria remaining on a surface is reduced in stainless steel, as compared to enamel, polycarbonate, and mineral resin surfaces after simulated wear tests.[2] For stainless steel surfaces used in areas where concerns for bacteria accumulation are paramount, polished surfaces should be considered.

Often in public restrooms stainless steel is used in urinal screens, toilet partitions, door handles, and countertops. Such fixtures are easy to clean using harsh chemicals, and they can be embossed to improve durability. The use of chlorinated detergents in areas where bacteria and viruses can proliferate is common. Repeated cleaning of the surfaces with these cleaners will not damage the stainless steel as long as concentrations of the cleaning agents do not remain on the surfaces or within seams. Bleach, a commonly used broad-spectrum bacteria killer, is a chloride compound that can have a corrosive effect on stainless steels. It is important to wipe the stainless steel down

[2] Kusumanigrum, H.D., Riboldi, G., Hazeleger, W.C., and Beumer, R.R. (2003). Survival of foodborne pathogens on stainless steel surfaces and cross-contamination to foods. *International Journal of Food Microbiology* 85 (3): 227–236.

afterwards to remove excess bleach. If bleach is allowed to remain on a stainless steel surface, it may cause it to corrode prematurely. If cleaning substances get into seams and stay moist, chlorine will attack the surface of the stainless steel and corrosion will occur. The seam is also where bacteria will collect, so good design should eliminate seams or incorporate methods of accessing the seams to clean the stainless steel thoroughly and eliminate pockets where bacteria can thrive or where chlorinated cleaners may become trapped and damage the integrity of the metal. Deionized water, also considered a bacterial killer, will clean stainless steel effectively and remove light fingerprints without chlorides. Deionized water is not as effective as bleach or other chlorine compounds, but as a final rinse, it can be an added deterrent and aid to remove chloride.

STAINLESS STEEL FOR THE ARTS

The art world has embraced stainless steel as a medium to produce some of the most remarkable sculptures ever created. Artists combine its durability and strength with the consistency of the surface finish. The artist lives in a visual world, and there are few other materials that interplay with light as well as stainless steel. In combining this with the inert behavior of the stainless steel surface, the artist can achieve a predictable, low-maintenance sculpture, regardless of exposure (Figure 1.10).

In much conventional sculpture, casting is a method used. The lost wax method of casting is as old as civilization itself and little has changed in the process. The metals used by the artist are often alloys of copper because of their ease of casting. These alloys have a rich history in art sculpture, and

FIGURE 1.10 The Petersen Automotive Museum, designed by Kohn Pedersen Fox.

the melting, flow, pattern development, welding, and patination are all well understood and have been in practice for centuries.

Stainless steel has different challenges, as it is newer to the art scene. There are stainless steel cast sculptures, but casting stainless requires equipment and kilns that can reach temperatures of 1500°C, significantly higher than temperatures for copper alloy castings. Finishing cast stainless steel is difficult and more time consuming than copper alloys. Milling operations may need to be introduced to arrive at a suitable surface. Stainless steel casting is also a method that requires a deep understanding of the metallurgical nature of steels. Because of these challenges, casting is not the most understood or available fabrication method for artists using stainless steel.

Stainless steel does offer the sculptor versatility in forms beyond casting. The potential lies in the mechanical strength offered by these special steels. Sculptures made from plates and assembled into large monoliths, such as the Gateway Arch in St. Louis, become a symbol of American ingenuity and display a form well fitted for the metal.

The tension/compression towers of Kenneth Snelson, which are assembled from stainless steel cable and polished rods of stainless steel, show another aspect where the mechanical properties of stainless steel coupled with the permanence of the surface offer possibilities that few other materials can provide (Figure 1.11).

Stainless steel can sparkle like sun reflecting on water or glow like a Bellini painting, in which light seemingly comes from the material itself. It is this nature of light reflection that attracts the artist to the metal (Figure 1.12).

FIGURE 1.11 *Triple Crown*, by Kenneth Snelson.

FIGURE 1.12 *Infinite Awareness*, by Jesse Small.

The edges of stainless steel, where light seems to concentrate, can define a shape. Whether cut with a laser into tight curves, or V-cut and folded, the edges will reflect the light as if they were drafted onto a surface. This phenomenon, involving light-wave propagation, is enhanced in metals because of the strong absorption and reemittance of light. The edges are where the reflected wave breaks down and the light at the edge no longer reflects on the same plane. Different surfaces of stainless steel with the same finish will appear to be different hues as the intensity changes, but the edge is the defining boundary. The edges of stainless steel reflect light slightly differently, as the interference of one reflected wave of light interacts with another. The edge can appear lighter, as if it were more polished, or darker, as if in a shadow. It is the reflective power of stainless steel that makes the edge more defining.

When stainless steel is cut, the edges capture the light and define the space. As the viewer moves, the image appears and disappears, as light reflecting off of the stainless steel edge seemingly turns on and off (Figure 1.13).

Stainless steel can be welded like the copper alloys, and the finish on the weld can be such that all indications of an assembly are concealed by this melding of materials. Even relatively thin plates and sheets can be assembled by welding with tight control and an understanding of the thermal properties of the metal. Distortions can be minimized or eliminated, and the subsequent grinding, polishing, and buffing can make the surface seamless (Figure 1.14).

Hand finishing can render a surface as intensely irregular as one in which gesso has been applied to a background, thus gaining the light/dark reflective changes that can liven the appearance as the

FIGURE 1.13 *Spannungsfeld*, by Julian Voss-Andreae.

FIGURE 1.14 *Cloud Gate*, by Anish Kapoor.

viewer moves within the near reflective range of the light. Away from the piece the disturbed surface becomes monolithic, as the reflections interfere with one another.

For the artist, the ability to finish and change the surface reflectivity of the metal or to utilize the strong reflective character stainless steel offers enables the sculptor to create with light. Light becomes the form itself. Consider the amazing sculpture by Walter De Maria, *The Lightning Field.* Here, Walter De Maria used 400 polished stainless steel rods spaced over a square mile in the desert of southwest New Mexico. The stainless steel rods capture the light. The viewer sees lines of light as the rods recede into the distance.

The artist Phillip K. Smith III also used polish stainless steel square tubes to create a stunning, 400-meter line of glowing pickets along the beach in Laguna, California (Figure 1.15).

Similar to other steels, stainless steel can be colored. Adding heat can induce colors in a less refined way by developing interference colors of blues, purples, reds, and even yellows. They are less refined because they are not as controllable or predictable and because they need to be further coated to retain their colors. Bluing or darkening stainless steels by chemically modifying the chromium on the surface of stainless steels is a process artists and craftspeople have perfected over the years and one that has been used to darken stainless steel for decades.

In more recent times, however, more refined and predictable coloring process have been developed and perfected. These also use interference coloring processes involving the controlled development of a thickened chromium oxide layer. Larger-scale projects can be produced using the process but there are limits, both dimensionally on a single element and on welding (Figure 1.16).

Another method of producing long lasting colors involves PVD. This method produces intriguing colors by applying a very thin (only a few molecules thick) coating of a compound, such as

FIGURE 1.15 *Quarter Mile Arc*, composed of polished stainless steel square tubes.
Source: Photo by Lance Gerber Photography, courtesy of Phillip K. Smith III.

FIGURE 1.16 *Butterflies*, by Ewardt Hilgemann.

titanium nitrate to produce gold or titanium carbide to produce a deep black. Again, there are limitations – but for an artist, it makes possible the introduction of color as well as finish to the sculpture. One of the first sculptures to incorporate colored stainless steel was *Ad Astra*, by Richard Lippold. This striking sculpture draws viewers "to the stars" with its sharp lines and mirrored reflective surface that rises over 30 m (100 ft.). Formed sheets of gold colored interference stainless steel are mixed with the natural color of stainless steel (Figure 1.17).

The permanence of stainless steel is a desirable quality. The fact that it does not have to be coated with lacquers or wax or paints or varnishes to keep it looking near to its original form is one of the attributes that make stainless a desirable material. Alexander Calder used stainless steel to create the massive *L'Homme* stabile form for the city of Montreal. Fabricated from 42 tn. of massive cut plates of stainless steel, this S30400 alloy form has stood since 1967. The surface has a coarse sandblast finish and has been thoroughly cleaned once. This massive sculpture, 28 m wide by 20 m tall, by one of the most famous sculptors of the twentieth century, celebrates stainless steel as a durable art form (Figure 1.18).

The available forms, finish possibilities, and color potential, and the interplay with light, make this metal an interesting option for artists and architects. Stainless steels expand the possibilities by the tremendous variety of forms, finishes, color, and alloys. Of all the metals used in art and architecture, stainless steel is the king.

FIGURE 1.17 *Ad Astra*, by Richard Lippold.

FIGURE 1.18 *L'Homme*, by Alexander Calder.
Source: Bohao Zhao. Used under CC BY 3.0, https://creativecommons.org/licenses/by/3.0.

CHAPTER 2

Alloys

"In that direction," the Cat said, waving its right paw round, "lives a Hatter: and in *that* direction," waving the other paw, "lives a March Hare. Visit either you like: they're both mad."

Lewis Carroll, *Alice's Adventures in Wonderland*

ALLOYS

In art and architecture, the choice of alloy is as much about aesthetics as corrosion and performance. A lot is made of choosing the "correct" alloy for a particular situation, when in reality many alloys would perform adequately for decades. Our built environment is cleaner, pollution is less of a concern, and the quality of the mill product is superior to that of the past. Salt environments—coastal and urban environments that use excessive deicing salts—are regions where alloy selection is critical. With good fabrication and design practice and a regularly scheduled cleaning regimen, most issues with corrosion will not present themselves. Stainless steel is one of the most durable, decorative surfacing materials ever created by mankind.

Stainless steel is an alloy of iron and chromium, with other elements added to enhance certain corrosion resistance behavior and industrial performance requirements. Gaining an understanding of the alloys of stainless steel is essential for determining the expected performance of the metal in various environments. Other than chromium, nickel and molybdenum are two important elements that have a beneficial effect on stainless steel.

For instance, nickel added to the iron and chromium mixture improves the corrosion resistance of the metal by stabilizing the chromium oxide layer as it develops on the surface. Small amounts of molybdenum, when added to the alloys with nickel, nitrogen, and chromium, further enhance the protective oxide layer by combatting the pitting behavior induced by the chloride ion.

Several other elements are added to make new stainless steel alloys. These typically make up less than 3% of the metal constituents, but they can influence the behavior in ways that are beneficial for a particular application or fabrication process. However, some of these additional elements may have ramifications in terms of appearance and corrosion resistance.

ALLOYING ELEMENTS

During the early development of stainless steel, particular attributes were sought in the metal that would complement the corrosion resistant characteristics. Attributes related to machining, heat strengthening, weldability, and improved corrosion resistance in more stringent environments were developed by adding various elements to the original alloys. Today there are dozens of specific alloys in use. Each has slightly a different composition that corresponds to the various alloys added to achieve a specific property.

Originally the alloys had names (such as V2A or Enduro S) created by the Mill to log the particular formula. In the United States, the alloys were later described with numbers, such as 302, 304, 304L, and 316. These numbers were established by American Iron and Steel Institute (AISI) and the Society of Automotive Engineers (SAE). In the 1930s and 1940s they sought to create a standard to describe these engineered alloys used in industry. In Europe they use another set of numerical values, defined by the European Committee for Standardization and beginning with EN for "European Standard." DIN Standards, as defined by the Deutsches Institut fur Normung in Germany, are also used. For example, the European stainless steel classifications for alloys 304 and 304L stainless steel alloys are 1.4301 and 1.4307, respectively. The different naming or numbering systems used to name the same stainless alloy perhaps falls under the heading of national pride, but it can make things confusing. In Appendix A there is a list of stainless steel alloys commonly used in art and architecture and a cross-reference for the various national standards. In this book we will be using the Unified Numbering System for Metals and Alloys (UNS) for designation of alloys. Granted, the UNS is a North American system managed by the American Society for Testing and Materials (ASTM) International and SAE International, but it has good logical grounding and is widely understood and accepted. The following sections provide a list of the most common alloying elements added to iron to make stainless steels.

Chromium

Corrosion resistance of stainless steel alloys improves as chromium content increases. However, production costs also increase and formability declines as more and more chromium is added. At least 10.5% chromium is needed for a metal to be classified as stainless steel. That is because with less than this level the protective oxide film will not develop. For example, the austenitic alloy S30400 contains 18–20% chromium.

Chromium gives stainless steel its distinctive luster.

Nickel

Nickel is needed to produce the austenitic crystal formation of the S3xxxx alloys. Nickel improves the strength and ductility of the austenitic stainless steels. Nickel aids the austenitic alloys in fending off corrosion from acid compounds. In S30400 stainless steel, there is between 8 and 10% nickel.

Molybdenum

Small percentages of molybdenum added to the chromium nickel alloys of stainless steel arrest the formation of pitting in the protective oxide layer.

Molybdenum-bearing stainless steels have excellent crevice corrosion resistance. Adding molybdenum will reduce the strength of the stainless steel. It will also change the luster of the chromium oxide layer.

Carbon

Carbon is in every stainless steel. It is what is added to iron to make steel. For stainless steels, though, it is desirable to have a very low level of carbon, at least for the austenitic and ferritic alloys. It is often purposely increased in martensitic alloys to increase the strength and increase the hardness of these alloys. But for austenitic alloys in particular, it is desirable to keep the carbon content low.

Manganese

The addition of this element helps stabilize the austenitic crystal formation when the metal is cast by preventing the development of iron sulfides. In the S2xxxx series of stainless steel alloys, it replaces some of the nickel.

Sulfur

The addition of sulfur to stainless steels aids in the machining of the metal; however, when these sulfides are exposed on the surface, they can become areas where corrosion initiates. Stainless steel pipe and rod often have additions of sulfur to facilitate further machining operations, such as threading and turning. When stainless steel is machined, the sulfur enables it to come off in small shards or chips. It also benefits rapid welding techniques. Sulfur reduces the heat that develops during some industrial manufacturing processes.

Sulfur will have a detrimental effect on the appearance and luster of stainless steel. Sulfur on the surface or near the surface makes polishing difficult and can leave areas where pitting corrosion can be initiated. For most applications, the sulfur content in the alloy should be less than 0.005%.

Selenium

Similarly to sulfur, selenium added to chromium nickel alloys improves machining. Selenium improves the cold working behavior of the metal as well. It is more expensive than sulfur, which is why sulfur is more commonly used.

Silicon

Silicon added to stainless steel has shown an improvement in the alloy's resistance to scale forming on the surface from repeated heating. For stainless steels subjected to high temperatures, such as exhaust systems, silicon is often added to resist what is known as carburization, a process in which the metal absorbs carbon from combustion or other carbon rich atmospheres.

In cast stainless steel, silicon improves fluidity.

Nitrogen

Small amounts of nitrogen are added to improve strength in the low-carbon alloys. In addition, nitrogen is added to duplex alloys to improve corrosion resistance. The addition of nitrogen together with molybdenum greatly improves pitting resistance.

Titanium and Niobium

Additions of titanium and niobium to stainless steel improve the resistance to carbide formation during welding operations. They have a greater affinity toward carbon and thus during the heat of welding they attract the carbon more readily than the chromium and stay dispersed throughout the metal and do not collect around the grain boundaries.

Copper, Osmium, and Tantalum

Copper as well as osmium and tantalum are sometimes added to stainless steels in very small amounts to improve corrosion resistance to acid or intergranular attacks in selected exposures.

Symbol	Element	Effect
Fe	Iron	Main element in steels and stainless steels
Cr	Chromium	Chromium provides corrosion resistance
Ni	Nickel	Nickel is needed for austenitic formation
Mo	Molybdenum	Added to improve corrosion resistance
C	Carbon	Decreases corrosion if present at grain boundaries
Mn	Manganese	Stabilizes austenitic alloys
S	Sulfur	Decreases corrosion resistance
Se	Selenium	Decreases corrosion resistance
Si	Silicon	With Mo present, improves corrosion resistance
N	Nitrogen	Improves corrosion resistance in the duplex alloys
Ti	Titanium	Reduces carbide precipitation
Nb	Niobium	Reduces carbide precipitation
Ta	Tantalum	Improves corrosion resistance in highly oxidized environments
Cu	Copper	Improves corrosion resistance to sulfuric acid

TEMPERS AND HEAT TREATMENT

When working with metals in art and architecture, you often hear the terms "heat treatment" and "precipitation hardening" in discussions of strength and mechanical characteristics. Stainless steels of the austenitic and martensitic families gain strength through a process referred to as precipitation hardening. This process heats the stainless steel to an intermediate temperature in the range of 650–760°C after it has been annealed in a temperature range of 1095–1120°C. The stainless steel is held for a period of time at this temperature, which allows small particles called "precipitates" to be dispersed throughout the alloy's crystal matrix structure. This increases the strength, because these precipitates that are now dispersed in the crystal structure resist dislocation as the stainless steel is stressed.

Most stainless steel alloys used in art and architecture are in the annealed state. Cold working stainless steel wrought forms will increase the strength; the vernacular terms used to describe the temper are ¼ hard, ½ hard, ¾ hard, and full hard. The mechanical information listed below each of the alloys in this chapter is an approximation of the mechanical properties of tensile strength, yield strength, and hardness for the annealed temper. The tensile and yield strength are given in KSI and MPa. KSI is the nonmetric unit of stress and stands for kips per square inch, where 1 kip is equal to 1000 pounds of force. MPa is the metric unit for force and stands for megapascals, defined as 1 million pascals. Hardness is described in terms of the Rockwell hardness scale or the Brinell hardness scale. Rockwell hardness is determined by the depth of a hardened intention into the material, whereas the Brinell test measures the diameter of the indention made by a load applied to a tungsten carbide ball. Both methods establish a hardness correlation with the temper of the metal. These should be verified with the supplier. An example using the approximate mechanical properties of the common austenitic alloy S30400 is shown in Table 2.1.

STAINLESS STEEL CLASSIFICATIONS

As noted in Chapter 1, there are four basic classifications of stainless steel:

- Austenitic
- Ferritic
- Martensitic (including the precipitation-hardened alloys)
- Duplex

TABLE 2.1 S30400 mechanical properties for various tempers.

Temper	Ultimate strength		Yield strength		Elongation % (2 in.)
	KSI	Mpa	KSI	Mpa	Min.
Annealed	90	621	42	290	55
1/4 hard	125	862	75	517	10
1/2 hard	150	1034	110	758	6
Full hard	185	1276	140	965	3

Each possesses certain characteristics derived from the crystal structure of the solid solution. Each classification has unique physical, magnetic, and corrosion properties. The austenitic, ferritic, and martensitic forms all were created prior to the 1930s and are the most common forms in use. The duplex alloys possess a combination of austenite and ferrite crystal structures and have made inroads into the architectural cladding industry because of their superior corrosion resistance in high-chloride exposures. But the austenitic form of stainless steel is still the prevalent stainless steel type used in architecture and art.

To describe stainless steel alloys, industry has adopted the UNS, which defines each particular alloy and its constituents. In the United States, AISI number classifications (such as 410, 302, 304, 316, etc.) are still in common use, but no new numbers have been generated since the 1980s. The UNS system is preferred because the specifics of an alloy's chemistry is more defined.

AUSTENITIC STAINLESS STEELS

The most commonly used alloys for architecture and art are the austenitic structure alloys. Austenitic stainless steel structure refers to a solid solution[1] possessing a face-centered cubic (FCC) structure. They make up over 70% of the stainless steel produced today. Austenitic stainless steels have a thermal expansion coefficient 50% higher than carbon steels. The strength of austenitic alloys cannot be increased by heat treatment processes, but it can be increased by cold working. They are annealed to maximize the corrosion resistance behavior that makes them so sought after as an architectural and artistic medium. Annealing temperatures of 1010–1121°C (1850–2050°F) dissolve the carbides and take them back into solution. Annealing also restores the ductility of these alloys.

Austenitic stainless steel can be divided into three categories:

1. *Unstabilized.* Examples are alloys S20100, S20200, and S30100 through S31700.
2. *Stabilized.* Examples are alloys S32100, S34700, and S34800.
3. *Extra low carbon.* Examples are alloys S30403 and S31603.

The difference between stabilized and unstabilized alloys is that chemical stabilizers, such as titanium and niobium, are added to the stabilized alloys. These elements, when added in minute amounts to the alloy mix, aid in the containment of carbon in critical heating and high temperature exposures. These are typically industrial alloys used in critical exposures.

Extra low carbon alloys are forms of the popular S30400 and S31600 alloys. Low carbon alloys are needed when extensive welding is to occur. With low carbon, there is less carbide precipitation during the heat of welding. This reduces chromium carbide precipitation and the associated reduction of corrosion resistance.

[1]"Solid solution" is a chemical term referring to a homogeneous crystalline structure of two or more atoms that share a crystal lattice structure.

The most common austenitic alloys in use today are those designated by AISI as 304 or 316, which are also known by the UNS alphanumeric designations S30400 and S31600, respectively.

Austenitic alloys are sometimes referred to as the 18-8 alloys of stainless steel because of the percentage of the element constituents of the main alloying components, chromium and nickel. The 18 is for a nominal 18% of chromium and the 8 refers to a nominal 8% of nickel in the steel alloy. This terminology is a carryover from past industry practice. Today it is wise to use the UNS designations. Specifying stainless steel as an 18-8 lacks preciseness. It should be noted that AISI created the values given for the names of the various alloys of stainless steel (with identifying numbers formatted as 302, 304, 316, etc.). These were created more than 40 years ago, and no new numbers have been added by AISI. The UNS system provides a much more detailed description of the alloying constituents.

Austenitic alloys have very good corrosion resistance, are readily shaped and formed, and possess the luster stainless steel is known for. The S31600 alloy provides improved corrosion resistance to chloride ion environments because of the addition of molybdenum. The molybdenum acts to stabilize the chromium oxide in the surface layer.

There are variations of these alloys that facilitate better welding behavior, such as S30403 and S31603. These are the alloys that are often referred to as low carbon (304L and 316L, respectively, under the AISI system). The L stands for low carbon in the older nomenclature. Carbon is maintained at less than 0.03% when it is cast and rolled or formed into sheet, plates, or other components at the Mill source. Most steel mills that manufacture stainless steel today provide low carbon as normal practice for thicknesses less than 3 mm (0.125 in.). It is still a good practice to specify this requirement if welding is to be performed.

The low carbon alloys provide the benefit of reduced chromium carbide formation at the recrystallization around welds. When steel is heated to what is called the critical range of 426–870°C, carbon precipitates out and combines with the chromium, forming chromium carbides along the grain boundaries. This weakens the area around the weld and exposes it to intergranular corrosion. Using the low carbon alloys when welding will aid in the reduction of this occurrence. It will also aid in the post-welding cleanup. The formation of chromium carbides is a function of time at the critical temperature range. The longer this area is at this critical temperature range, the more chromium carbide will be formed around the grain boundaries in the areas near the weld. The greater the thickness of the parts being welded, the longer the area is at this critical temperature range.

Some Mills purchase their base steels in low carbon grades, and thus you will often see mill certificates that say 304/304L as the grade. "Grade" is a term used by industry, but it does not represent any particular quality aspect that differentiates one alloy from the other. If it is used in that sense it is misleading. Carbon does not influence the corrosion resistance behavior of stainless steel or the appearance of the metal except when the metal is welded, and then only at the region around the weld.

The low carbon forms are known by the UNS descriptors S30403 and S31603; common forms are shown in Figure 2.1.

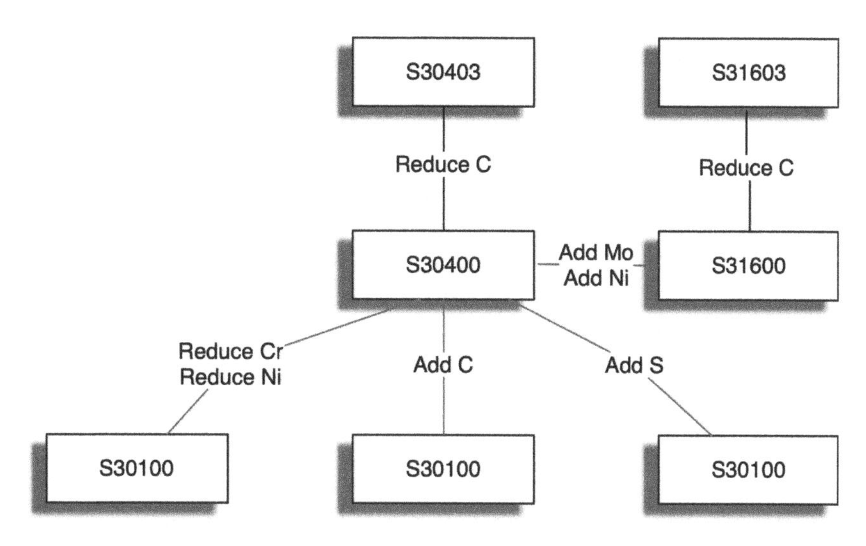

FIGURE 2.1 Common austenitic alloys.

MAGNETISM AND MAGNETIC PERMEABILITY

Austenitic stainless steel can undergo changes in its atomic structure to impart ferromagnetic characteristics to the metal. Under certain cold working operations, such as deep drawing and bending, the metal shifts locally from an FCC structure to a body-centered tetragonal structure.

Using magnetism as a test for alloy type can be misleading. In the annealed state, which is the state of most flat sheets and plates, austenitic S3xxxx series stainless steel alloys will show no magnetic permeability; that is, no magnetism. Once the sheet undergoes extensive forming, deep drawing, and even abrasive finishing, the surface will exhibit a magnetic response as the crystal structure alters to the body-centered structure corresponding to martensite.

There is no correlation between corrosion resistance and magnetism when it comes to the austenitic grades of stainless steel. Even S31600 alloys of stainless steel can show magnetic attraction when they are cold forged or deeply drawn.

On deep drawn objects, the magnetism can vary across the structure. Where the metal has undergone significant cold working, the surface is magnetic; where it has not experienced the cold work, it is not magnetic. Extensive finishing of the surface, blasting, or even deep directional sanding increases the magnetic characteristic of the stainless steel surface.

For fasteners, S3xxxx series austenitic stainless steel can exhibit magnetism. The fastener heads are created by cold forging the shapes and can exhibit magnetism.

Cast grades of austenitic stainless steel (J92500, for example) do not have precisely the same chemistry as their wrought version (S30400 in this example). The cast versions have a different metallurgic structures with different levels of ferritic crystal structures intermixed. Since ferritic crystallization can form along with the austenitic crystal structure, castings often exhibit magnetism.

FERRITIC STAINLESS STEELS

Ferritic stainless steel is a body-centered cubic (BCC) crystal structure. Ferritic alloys of stainless steel do exhibit magnetic permeability. The ferritic stainless steels are often used in food service because of their resistance to the organic acids found in food and because they are more economical due to the reduction of nickel in their alloying makeup. Ferritic stainless steels possess chromium, similarly to austenitic stainless steel. They will show signs of rust when exposed to most outdoor climates. The rust is more of an aesthetic concern and can be removed. Ferritic stainless steels are typically used where they will be kept clean, such as in kitchens and other food preparation areas. They are also used for automotive trim.

Some of the common alloys of the ferritic stainless steel are shown in the following table and in Figure 2.2.

S43000	Used in food service
S43400	Used as automotive trim and for some fasteners
S44200	Widely used precipitation-hardened alloy for industry

MARTENSITIC STAINLESS STEELS

The third form is martensitic stainless steel. Martensitic stainless steel possesses a tetragonal crystal structure formed by the rapid cooling or quenching of austenitic stainless steel. Here the FCC structure is transformed rapidly into a highly strained body-centered tetragonal crystal.

Martensitic stainless steel has good corrosion resistance, similar to the ferritic stainless steels, and they are also magnetic. Knives, tools, and fasteners are created from this alloy due to its increased hardness and its ability to hold an edge. Common alloys of martensitic stainless steels include those shown in the following table and in Figure 2.3.

S41000	Used for fasteners, widely used in industry
S41400	Used for knives
S41600	Used for fasteners
S42000	Used for surgical equipment

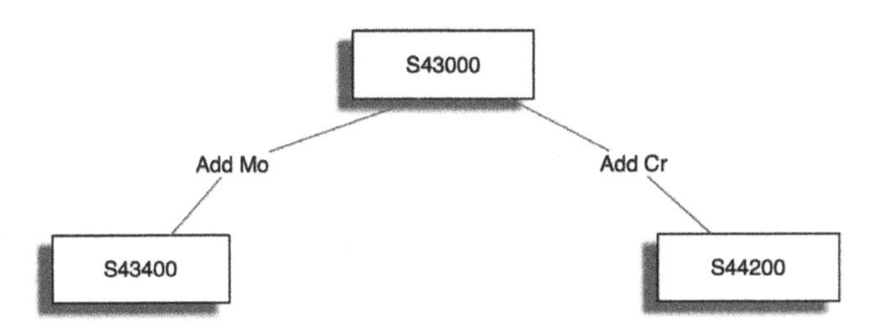

FIGURE 2.2 Common ferritic alloys.

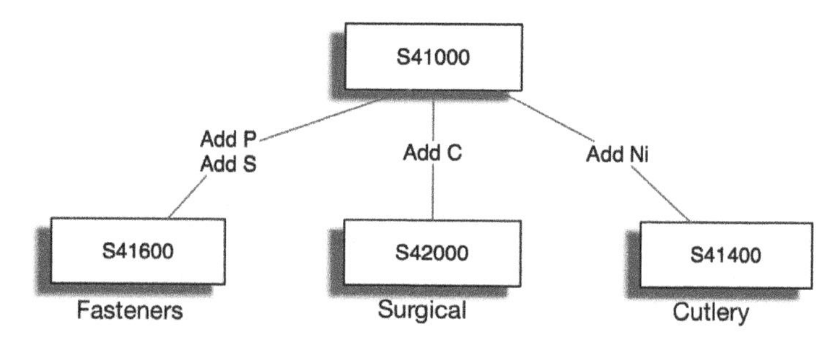

FIGURE 2.3 Common martensitic alloys.

PRECIPITATION-HARDENED STAINLESS STEEL

There is also a form of stainless steel called "precipitation-hardened." Used often for bolts in assemblies, the most common form is 17-4PH, designated as S17400 under the UNS system (Figure 2.4). This is a martensitic form of stainless steel; it possesses corrosion characteristics similar to austenitic stainless steel.

Precipitation-hardened stainless steels have tensile and yield strengths several times that of the austenitic stainless steels. This, coupled with their good corrosion resistance, makes them ideal for bolts. Low temperature heat treatment processes can improve strength and hardness. Ferritic alloys can also be precipitation hardened to improve strength.

FIGURE 2.4 Precipitation-hardened stainless steel bolt with a 17-4 marking (bolt is shown without the PH stamp).

DUPLEX STAINLESS STEELS

Duplex alloys of stainless steel are growing in use on architectural projects.

Duplex alloys are used in areas where salt environments are prevalent and cleaning processes are minimum or nonexistent. The duplex alloys are a metallurgic combination of the austenitic and the ferritic forms of stainless steel. The ratio is approximately 60:40 austenite to ferrite. They have improved strength and corrosion resistance and reduced amounts of nickel. The duplex alloys have a higher chromium content, and this, coupled with molybdenum and nitrogen, results in an alloy that resists pitting behaviors in the metal when exposures to chlorides are prevalent. Nickel is the additive element that forms the austenite structure, while chromium and molybdenum form the ferritic structure.

These alloys came into existence with modern steelmaking technology. They were originally developed in Sweden back in 1930, but they were difficult to manufacture. New techniques enabled steel manufactures to have tighter controls on the nitrogen. They were introduced commercially in the late 1970s.

Duplex stainless steel has a low carbon content (less than 0.03%). Duplex alloys are significantly stronger—several times the yield strength of austenitic alloys—but they are less ductile than standard austenitic alloys. The dual grain structure makes these alloys harder to cold work as compared to austenitic and ferritic stainless steel (Figure 2.5). They are stiffer and exhibit a greater amount of springback when formed.

Conventional forming equipment can overcome the higher yield strength of duplex stainless steel, but its inelastic nature poses challenges for shaping and forming. Roll forming, in particular, requires adjustments for the differences in elasticity of the duplex stainless steels. Due to their ferritic structure, they are magnetic. Additionally, they have a higher heat conductivity than austenitic stainless steels and a lower thermal expansion coefficient.

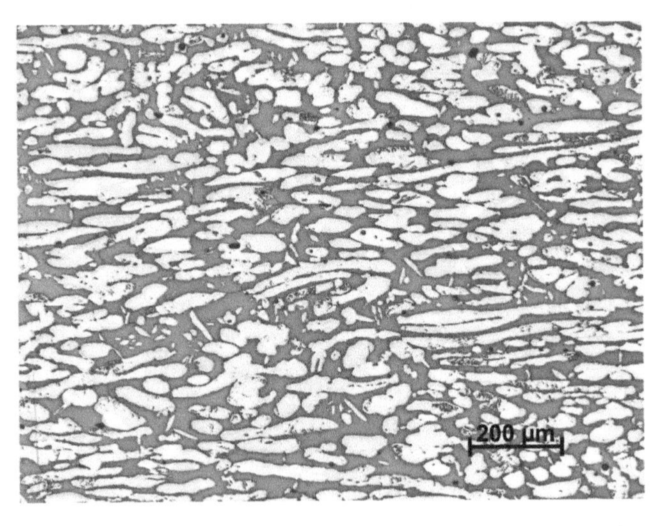

FIGURE 2.5 The microstructure of duplex stainless steel shows both austenitic and ferritic structure.

Color and finish are similar to those of austenitic stainless steels. Finishing and texturing processes are interchangeable with those used on austenitic alloys.

There are several alloys of duplex stainless steel in use, which are listed by their UNS designations in the three lists that follow. The S31803 alloy (also known as 2205 alloy) is the most commonly used duplex alloy and makes up over 80% of the worldwide production. The super duplex alloys are rarely if ever used in art and architecture. There are even so-called hyper duplex stainless steel alloys that have as much as 33% chromium.

Lean Duplex Alloys (Lower Alloying Constituents)

S32101

S32202

S32304

S32002

Standard Duplex Alloys (19.5–23% Chromium)

S31803/S32205 (also referred to as 2205 alloy)

Super Duplex Alloys (24–30% Chromium)

S32760

S32750

CHOOSING THE RIGHT ALLOY

The choice of the correct stainless steel alloy is dependent on a number of variables. We can assume that the goals are to obtain a surface or component that will last at least a lifetime, will have predictable appearance behavior, and will need to undergo cleaning infrequently beyond that which is provided by nature in the form of occasional rain.

The variables to consider in the selection of a stainless steel alloy follow, while Figure 2.6 provides a decision tree for choosing alloys compatible with given conditions.

- Proximity to the coast
- Exposure to deicing salts
- Humidity
- Cleaning regimen
- Welding requirements
- Forming requirements
- Machining requirements
- Wrought or cast form
- Strength requirements

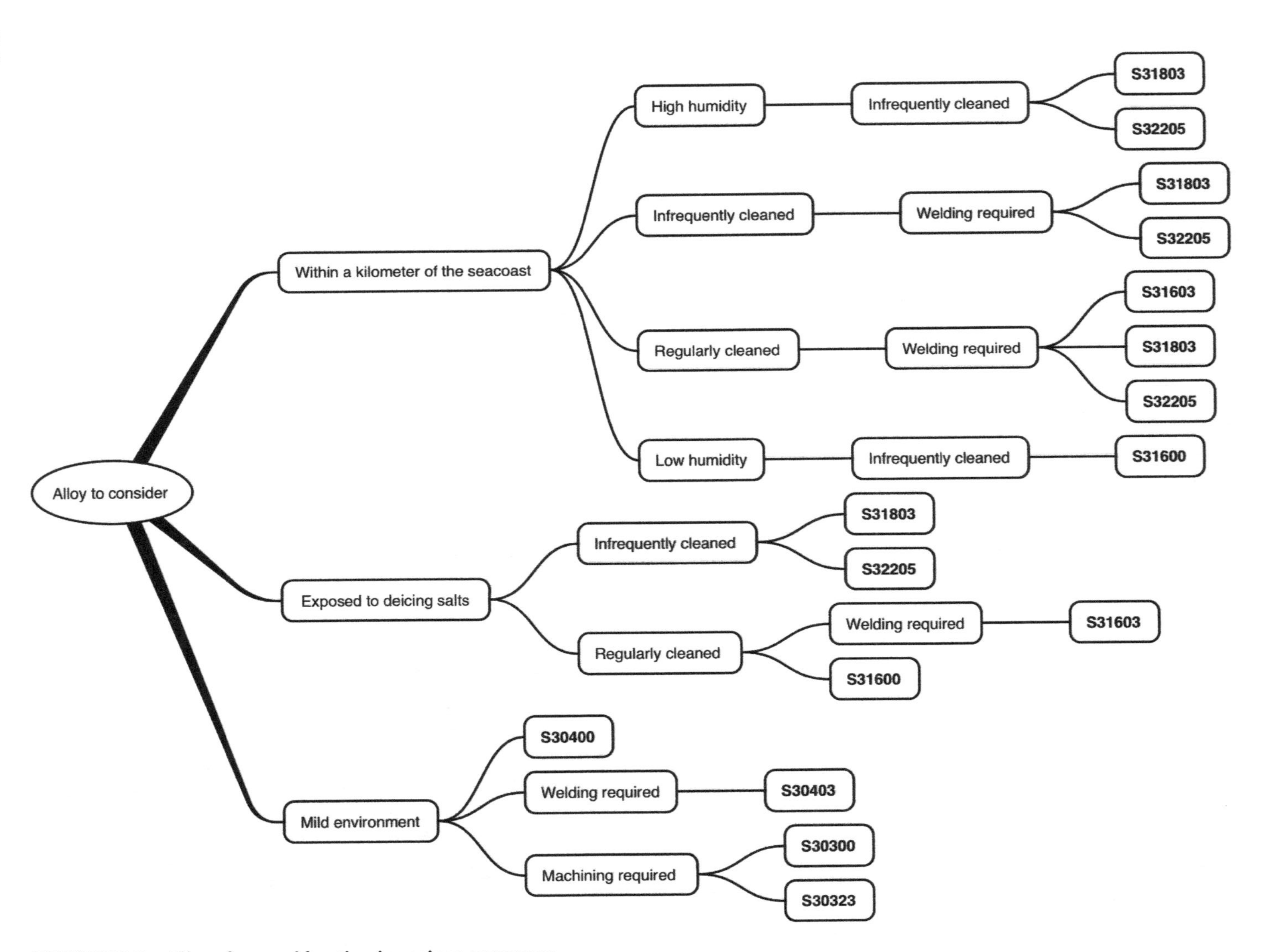

FIGURE 2.6 Alloys for consideration in various exposures.

TABLE 2.2 Pitting resistance equivalent number (PREN) of various alloys.

Alloy	Structure	PREN
S43000	Ferritic	18
S30400	Austenitic	19
S43932	Ferritic	19
S32001	Duplex	22
S31600	Austenitic	24
S44400	Ferritic	24
S31603	Austenitic	26
S32101	Duplex	26
S31703	Austenitic	28
S32205	Duplex	35
S31254	Austenitic	44

In art and architecture, where appearance usually eclipses all other concerns, how the metal will perform in a given exposure is paramount. Once stainless steel becomes the metal of choice for a particular application, the cost difference among alloys plays only a small role; the labor involved in fabrication and installation is the greater cost. Therefore, choice of alloy will be dictated mostly by the environmental exposure. For stainless steel, a humid, chloride ion environment is the most detrimental. Chloride ions will initiate corrosive attacks on stainless steel, which can lead to pitting corrosion.

The stainless steel industry has devised a simple algorithm to act as a guide for ranking how a stainless steel alloy may perform in chloride ion exposures. It is called the "pitting resistance equivalent number," or "PREN" (Table 2.2). This algorithm takes into consideration the main corrosion resistant alloying constituents and their weighted benefit in preventing pitting corrosion. As with most weighted formulas, it is not an absolute determination but a statistically derived comparative evaluation.

The formula is as follows:

$$\text{PREN} = (\%\ \text{chromium}) + 3.3 \times (\%\ \text{molybdenum}) + 16 \times (\%\ \text{nitrogen})$$

Pitting resistance increases as more molybdenum and nitrogen are introduced. The highest PREN value, alloy S31254, is not a common architectural alloy because of its cost and the difficulty in working it. It is known as a super austenitic alloy. S31254 is used in places where exposure to chlorides is extremely high.

Another factor that factors into the choice of stainless steel alloy is cost. The most common alloy used in art and architecture is S30400, followed by S31600. In fact, over 95% of the alloys used in art and architecture around the world are one of these alloys (or their low carbon sisters). The alloy S30400 is the workhorse and is also one of the most economical of the alloys. The trade-off is in the corrosion resistance in certain atmospheres. Even with the lower PREN value, if the surfaces are kept clean, these alloys will perform well. So in the end it boils down to how rigorous, thorough, and frequent the cleaning process can be. The reality is that these stainless steel surfaces will not be cleaned as often as an automobile or even a glass curtain wall. The metal will be assumed to last and perform; therefore a more specialized alloy may be needed to protect the surfaces from certain atmospheres when cleaning is infrequent.

Still, for stainless steel to perform as designed, it does require some level of maintenance. The Chrysler Building roof has undergone cleaning once since it was installed. There are many other structures that have only been cleaned by nature so far. The Gateway Arch has never been cleaned except by natural means. Many large-scale stainless steel roofs have not seen cleaning except by natural means and they still look incredible and perform well.

The following is a list of alloys with certain attributes that fall into the art and architecture realm of use. Some are more suited for certain environments and are identified for a particular industry or application. The list begins with the wrought version of the alloy. Wrought stands for the non-cast forms that have undergone some level of cold or hot work as opposed to cast stainless steel which has not been subjected to hot or cold working. The grain behavior and structure of the grains is different.

In wrought forms of stainless steel, we consider sheet, plate, bar, rod, tube, pipe, and wire forms. The specific alloys relate to these forms of stainless steel.

ARCHITECTURAL ALLOYS: WROUGHT SHEET, PLATE, BAR, ROD, PIPE, TUBE, AND WIRE

Ferritic Alloys

The first group of alloys is the ferritic family. The ferritic alloys of stainless steel are characterized by their BCC crystal structure. This structure, like that of iron, gives it the property of magnetism. These alloys are not commonly specified in art and architecture due to their lower corrosion resistance. These alloys do not weld as effectively as the austenitic family and can crack around the welds. Generally they are more economical than other stainless steel forms because they have low amounts of nickel. These alloys are used in kitchens and food preparation areas, breweries, and for automotive trim. They have good ductility and can be finished in similar ways to the austenitic alloys.

Listed are only a few of the ferritic alloys in use. There are many more custom alloys designed for specific applications and exposures.

STAINLESS STEEL ALLOY S43000

Ferritic

UNS S43000	ASTM A240 ASTM A480/A
430 EN 1.4016 SUS 430	

Alloying Constituents Added to Iron (wt. percentage)

C	Cr	Ni	Mn	Si	P	S
0.12% max	16–18%	0.75% max	1% max	1% max	0.040% max	0.03% max

	Tensile strength		Yield strength		Elongation % (2 in.–50 mm)	Rockwell hardness
S43000	70 KSI	483 MPa	45 KSI	310 MPa	22	85

Available Forms

Foil

Sheet

Plate

Bar

Tube

Bolts

Stainless steel alloy S43000 is a common ferritic alloy of stainless steel. It is the stainless steel predominately used in food service and is sometimes referred to as food grade stainless steel. It is also used in automotive trim and in appliances. It has decent corrosion resistance, particularly when it receives regular cleaning. It should not be used in chloride environments.

STAINLESS STEEL ALLOY S43900

Ferritic

UNS S43900	ASTM A959 ASTM A276 ASTM A240 ASTM A632
439LT EN 1.4509 SUS 430LX	

Alloying Constituents Added to Iron (wt. percentage)

0.03% max 17–19% 0.75% max 1% max 1% max 0.03% max 0.040% max 0.03% max

	Tensile strength		Yield strength		Elongation % (2 in.–50 mm)	Rockwell hardness
S43900	62 KSI	430 MPa	36 KSI	250 MPa	18	—

Available Forms

Plate

Bar

Tube

Pipe

This wrought ferritic alloy is very similar to S30400 stainless steel as far as corrosion resistance. It is used for some exterior surfacing applications and exhaust systems and is not a common architectural alloy. It is included because of its good corrosion characteristics.

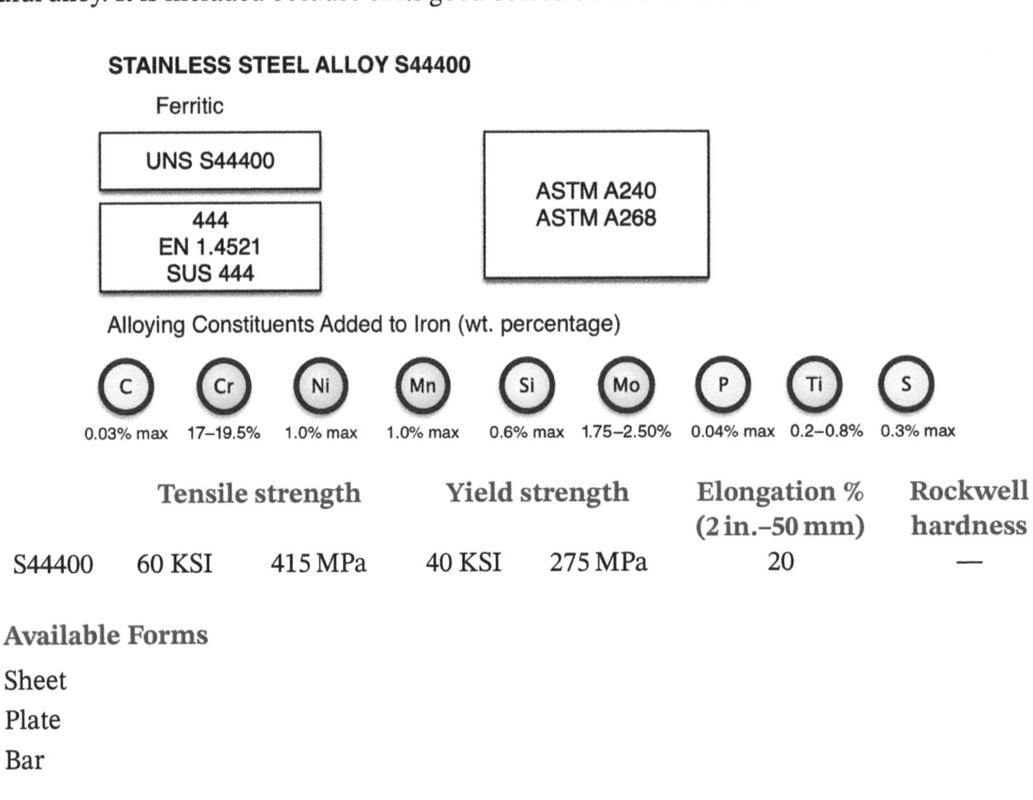

	Tensile strength		Yield strength		Elongation % (2 in.–50 mm)	Rockwell hardness
S44400	60 KSI	415 MPa	40 KSI	275 MPa	20	—

Available Forms

Sheet

Plate

Bar

Alloy S44400 is a corrosion resistant ferritic alloy. It is commonly used in food processing, wine, and beer brewery applications. It has good corrosion resistance.

Martensitic Alloys

These are not common in art and architecture except as bolts and fasteners. They have moderate corrosion resistance but good strength. Listed here are the only two alloys that are used (infrequently) in art and architecture. These alloys are magnetic.

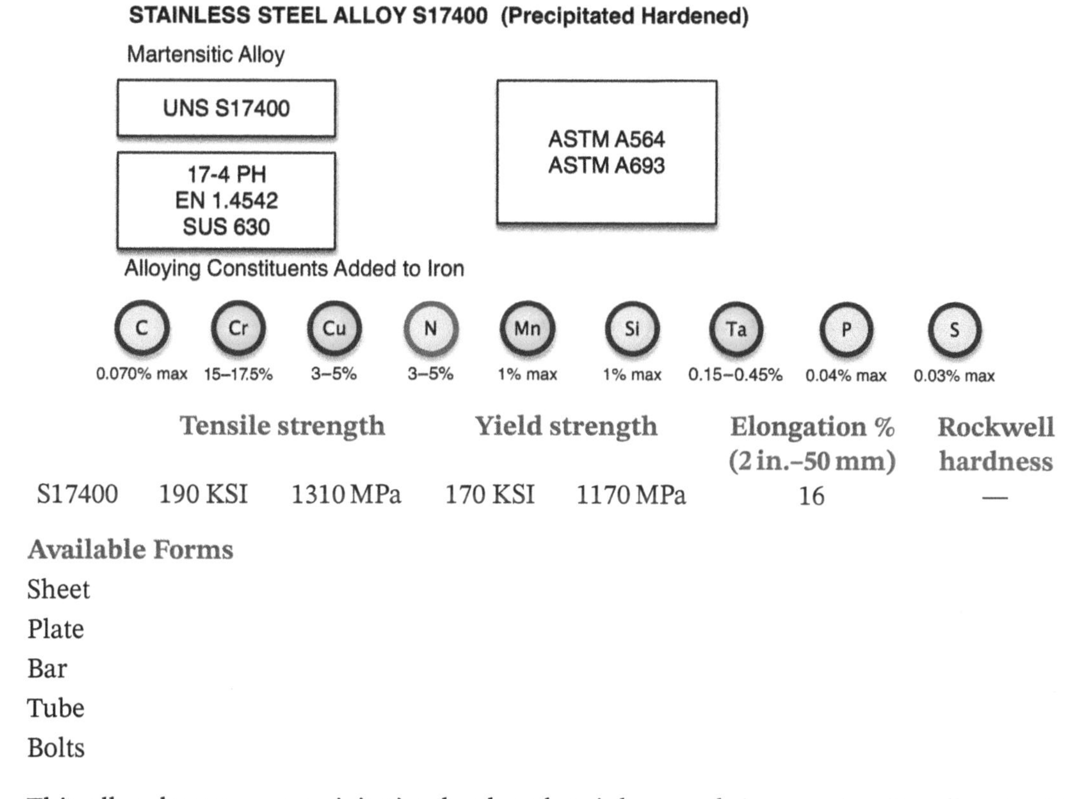

	Tensile strength		Yield strength		Elongation % (2 in.–50 mm)	Rockwell hardness
S17400	190 KSI	1310 MPa	170 KSI	1170 MPa	16	—

Available Forms

Sheet

Plate

Bar

Tube

Bolts

This alloy, known as precipitation-hardened stainless steel, is a martensitic alloy. Architecturally, it is used most often in stainless steel bolts due to its high tensile and yield strength. The bolts are usually stamped with "PH" to indicate their makeup. The corrosion resistance of the S17400 alloy is only moderate as compared to the other alloys.

STAINLESS STEEL ALLOY S41000

Martensitic

UNS S41000	ASTM A240
410 EN 1.4006 SUS 410	ASTM A480/A

Alloying Constituents Added to Iron (wt. percentage)

0.15% max

11.5–13.5%

0.50% max

1% max

1% max

0.040% max

0.03% max

	Tensile strength		Yield strength		Elongation % (2 in.–50 mm)	Rockwell hardness
S41000	65 KSI	450 MPa	30 KSI	205 MPa	20	96

Available Forms

Sheet

Plate

Bar

Tube

Fasteners

Alloy S41000 is the more common of the martensitic alloys. It is used for making knives because it holds an edge well. It is also used in fasteners and bolts.

STAINLESS STEEL ALLOY S42000

Martensitic

UNS S42000

420
EN 1.4021
SUS 420J1

ASTM A240
ASTM A480/A

Alloying Constituents Added to Iron (wt. percentage)

C	Cr	Ni	Mn	Si	P	S
0.15% max	12–14%	0%	1.25% max	1% max	0.040% max	0.03% max

	Tensile strength		Yield strength		Elongation % (2 in.–50 mm)	Rockwell hardness
S42000	95 KSI	655 MPa	50 KSI	345 MPa	25	—

Available Forms

Sheet

Plate

Bar

Fasteners

This alloy is a variation of alloy S41000. Its main attribute is that it can be hardened to 50 HRC (Hardness Rockwell Scale C). Knives and shear blades used to cut other metals are made from this alloy. It has good corrosion resistance if cleaned regularly.

Austenitic Alloys

The austenitic alloys make up the greatest share of stainless steel alloys used for art and architecture. The austenitic alloys are nonmagnetic in the annealed condition but will become partially magnetic when cold worked. These alloys have an FCC structure due in part to the alloying elements nickel, manganese, and nitrogen. The alloys offer excellent corrosion resistance in most environments. They are ductile and can be welded with all welding processes.

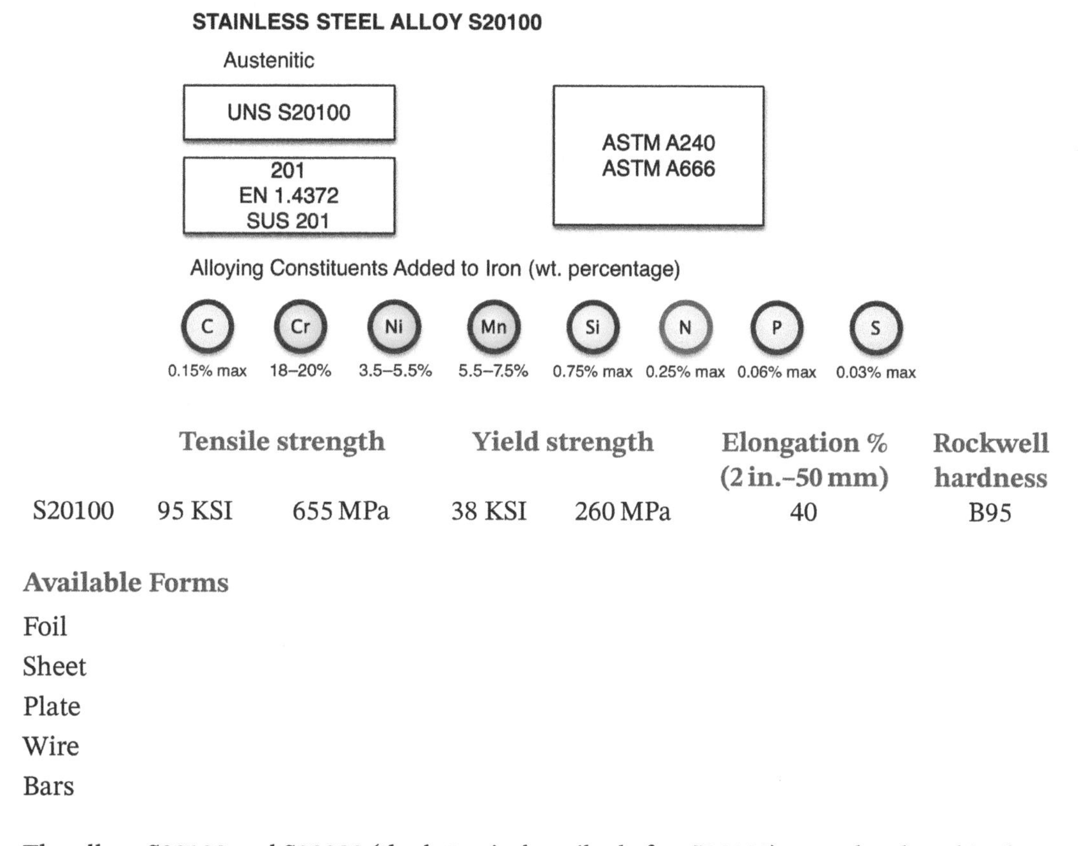

	Tensile strength		Yield strength		Elongation % (2 in.–50 mm)	Rockwell hardness
S20100	95 KSI	655 MPa	38 KSI	260 MPa	40	B95

Available Forms

Foil
Sheet
Plate
Wire
Bars

The alloys S20100 and S20200 (the latter is described after S20100) were developed in the 1950s when nickel was scarce and its price escalated. It was considered an alternative to the S30200 alloy. The corrosion resistance is similar to S30100, with a PREN value of around 20. Usually given a cold rolled temper that increases the yield strength, it is used for spring metal as well as food service equipment, in automotive-related uses, and for kitchen utensils.

STAINLESS STEEL ALLOY S20200

Austenitic

UNS S20200	ASTM A240
202 EN 1.4373	ASTM A666

Alloying Constituents Added to Iron (wt. percentage)

0.15% max

17–19%

4.0–6.0%

7.5–10.0%

1.0% max

0.25% max

0.06% max

0.03% max

	Tensile strength		Yield strength		Elongation % (2 in.–50 mm)	Rockwell hardness
S20200	100 KSI	700 MPa	45 KSI	310 MPa	14–45	—

Available Forms

Sheet

Plate

Bar

Tube

The alloy S20200 is similar to S20100 but has higher yield strength and tensile strength. Tempering increases this further. This is a relatively low-cost austenitic alloy that has moderate corrosion resistance. Its PREN value is 20.

STAINLESS STEEL ALLOY S30300

Austenitic

UNS S30300	ASTM A194 ASTM A895
303 EN 1.4305 SUS 303	ASTM A314 ASTM A473

Alloying Constituents Added to Iron (wt. percentage)

0.15% max

17–19%

8–10%

2% max

1% max

0.2% max

0.15–0.35%

STAINLESS STEEL ALLOY S30323

Austenitic

UNS S30323	ASTM A194 ASTM A895 ASTM A314 ASTM A473
303Se EN 1.4325	

Alloying Constituents Added to Iron (wt. percentage)

C	Cr	Ni	Mn	Si	P	S	Se
0.15% max	17–19%	8–10%	2% max	1% max	0.2% max	0.06% max.	0.15–0.35%

	Tensile strength		Yield strength		Elongation % (2 in.–50 mm)	Rockwell hardness
S30300	75 KSI	517 MPa	30 KSI	207 MPa	35	76
S30323	90 KSI	620 MPa	35 KSI	240 MPa	50	76

Available Forms

Sheet

Plate

Bar

Tube

These alloys are not as corrosion resistant as S30400, but the addition of sulfur and selenium improves the machining capability of these stainless steels. These alloys are used for fittings, valves, fasteners, and other highly machined stainless steel parts.

STAINLESS STEEL ALLOY S30400

Austenitic

UNS S30400	ASTM A240 ASTM A666
304, 302 EN 1.4301 SUS 304	

Alloying Constituents Added to Iron (wt. percentage)

C	Cr	Ni	Mn	Si	N	P	S
0.08% max	18–20%	8–10.5%	2% max	1% max	0.1% max	0.045% max	0.03% max

STAINLESS STEEL ALLOY S30403

Austenitic

UNS S30403	ASTM A240 ASTM A666
304L EN 1.4307 SUS 304LN	

Alloying Constituents Added to Iron (wt. percentage)

C	Cr	Ni	Mn	Si	N	P	S
0.03% max	18–20%	8–12%	2% max	1% max	0.1% max	0.045% max	0.03% max

	Ultimate strength		Yield strength		Elongation % (2 in.–50 mm)	Rockwell hardness
S30400	90 KSI	621 MPa	42 KSI	290 MPa	55	82
S30403	85 KSI	586 MPa	35 KSI	241 MPa	55	80

Available Forms

Sheet

Plate

Bar

Tube

Pipe

Wire

S30400 and S30403 are the workhorse alloys of stainless steel. Alloys S30400 and S30403 are the two most commonly used alloys in art and architecture. These alloys have excellent corrosion resistance in most applications and exhibit good visual clarity and luster. For most architectural uses it is advisable to keep the sulfur content as low as possible. Sulfur on the surface will have an undesirable effect on the luster and can create small surface inclusions when polished.

S30403 is the low carbon alloy. Today Mills work to keep the carbon low in thin sheet (less than 3 mm). In production, S30400 may be the low carbon version by default, and often the mill certification verifying the alloy shows both S30403 and S30400 as 304/304L. Alloy S30403 has slightly lower strength due to the reduction of carbon. These alloys are available in both wrought and cast forms. They have good workability and are available in various cold rolled tempers.

STAINLESS STEEL ALLOY S31600

Austenitic

UNS S31600	ASTM A182 ASTM A276 ASTM A479
316 EN 1.4401 SUS 316	

Alloying Constituents Added to Iron (wt. percentage)

C	Cr	Ni	Mo	Mn	Si	N	P	S
0.08% max	16–18%	10–14%	2–3%	2% max	1% max	0.1% max	0.045% max	0.03% max

STAINLESS STEEL ALLOY S31603

Austenitic

UNS S31603	ASTM A182 ASTM A276 ASTM A479
316L EN 1.4404 SUS 316LN	

Alloying Constituents Added to Iron (wt. percentage)

C	Cr	Ni	Mo	Mn	Si	N	P	S
0.03% max	16–18%	10–14%	2–3%	2% max	1% max	0.1% max	0.045% max	0.03% max

	Tensile strength		Yield strength		Elongation % (2 in.–50 mm)	Rockwell hardness
S31600	75 KSI	515 MPa	30 KSI	205 MPa	40	95
S31603	70 KSI	486 MPa	25 KSI	170 MPa	40	95

Available Forms

Sheet

Plate

Bar

Tube

Pipe

Wire

Alloys S31600 and S31603 are the second-most common alloys used in art and architecture. They possess a slightly elevated chromium content and have molybdenum added. The molybdenum benefits the corrosion resistance of the metal by helping to stabilize the chromium oxide layer when pitting is initiated by chloride ions. Alloys S31600 and S31603 are exceptional corrosion resistant stainless alloys and will perform well when used in moderately corrosive environments, such as coastal regions, and for surfaces prone to being inflicted with deicing salt residue. Alloy S31603 is a low carbon austenitic alloy of stainless steel. This alloy has superior corrosion and good pitting resistance. Also known as 316L, the low carbon content of S31603 helps to maintain good corrosion resistance when welded or heat formed. Post-welding cleanup will be more limited due to the reduction of chromium carbide precipitation around the weld. These alloys have shown extremely good performance if occasionally cleaned with fresh water.

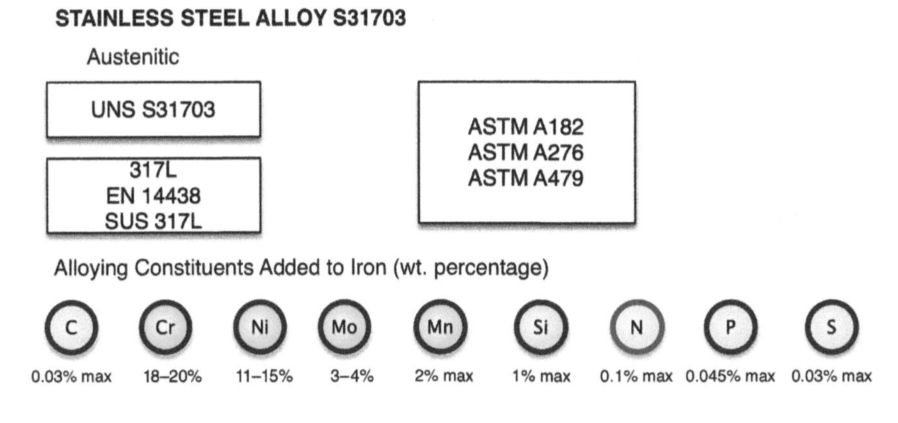

	Tensile strength		Yield strength		Elongation % (2 in.–50 mm)	Rockwell hardness
S31703	75 KSI	515 MPa	30 KSI	205 MPa	40	96

Available Forms

Sheet

Plate

Bar

Tube

Wire

Alloy S31703 is not a common architectural alloy. It is used in highly corrosive environments, such as paper plants and chemical and petrochemical facilities. This alloy has a PREN value of 30, which makes it very effective in pitting environments. The increased chromium and nickel make this alloy exceptional in many highly corrosive environments. It is considered a high-performance austenitic stainless steel alloy.

STAINLESS STEEL ALLOY S31254

Austenitic

UNS S31254	ASTM A240 ASTM A813 ASTM A469 ASTM A269
EN 1.4547 SUS 312L	

Alloying Constituents Added to Iron (wt. percentage)

0.02% max	19.5–20.5%	17.5–18.5%	1% max	0.8% max	6–6.5%	0.3% max	0.1% max	0.5–1.0%

	Tensile Strength		Yield Strength		Elongation % (2 in.–50 mm)	Rockwell hardness
S31254	99 KSI	680 MPa	43 KSI	300 MPa	50	—

Available Forms

Sheet

Plate

Bar

Tube

Pipe

Alloy S31254 is what has been referred to as a "super austenitic" stainless steel. It is difficult to machine due to its high strength and toughness. Not normally considered for art or architecture, this austenitic alloy possesses one of the highest PREN values: 44. It is used in severe corrosion environments where aqueous solutions containing chlorides, such as in desalination plants and with bleaching equipment, are prevalent.

This alloy is very difficult to cold form because it has a high work-hardening rate. Cold working will increase the strength and toughness, but more energy is required to shape it.

Duplex Alloys

These alloys, known as duplex stainless steel, have a significantly higher yield that other alloys of stainless steel. Their strength and hardness are both a benefit and detriment in use. They possess a biphase structure of austenitic and ferritic crystals; thus the name "duplex." They have been shown to have a better corrosion resistance to humid, high-chloride environments than the common austenitic alloys of S30400 and S31600. For all intents and purposes, the appearance of duplex stainless steels is similar to that of S30400 stainless steel.

They are magnetic due to the ferritic portion of the grain structure. They are stronger and tougher than the austenitic family of alloys. Duplex alloys are more difficult to work due to their higher strength, and they exhibit a greater springback character than the other steels. Forming with conventional equipment will take more power and a better understanding of the behavior of this metal. These alloys can be welded with conventional welding equipment, but the process is slightly different due to the nature of the duplex form. The following are the more common duplex alloys used in art and architectural applications where a corrosion resistant alloy is required.

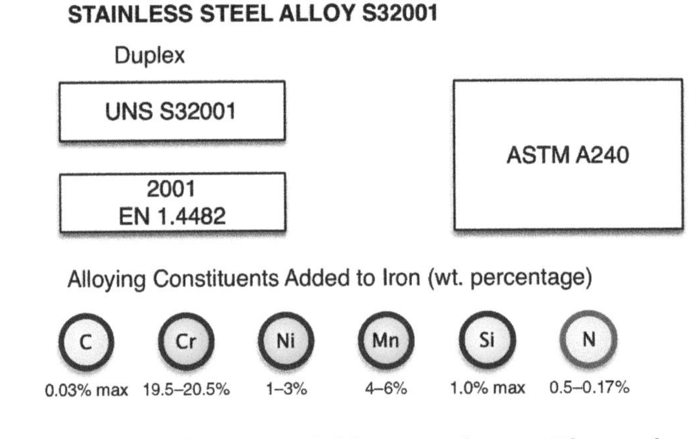

	Tensile strength		Yield strength		Elongation % (2 in.–50 mm)	Rockwell hardness
S32001	90 KSI	621 MPa	65 KSI	448 MPa	25	—

Available Forms

Sheet
Plate
Rod
Tube
Pipe

This duplex alloy of stainless steel has good corrosion resistance and improved weld capability. Its strength and toughness are similar to other duplex alloys used in architecture.

STAINLESS STEEL ALLOY S32101

Duplex

UNS S32101	ASTM A240
2101 EN 1.4162	

Alloying Constituents Added to Iron (wt. percentage)

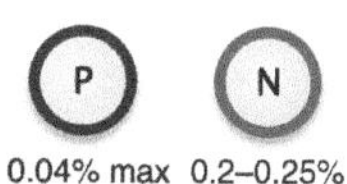

C	Cr	Ni	Mn	Si	Mo	P	N	Cu
0.03% max	21–22%	1.35–1.70%	4–6%	1% max	0.1–0.8%	0.04% max	0.2–0.25%	0.1–0.8%

	Tensile strength		Yield strength		Elongation % (2 in.–50 mm)	Rockwell hardness
S32101	101 KSI	696 MPa	65 KSI	450 MPa	38	—

Available Forms

Sheet

Plate

Bar

Tube

Pipe

This duplex alloy is very difficult to work due to its very high strength. S32101 has superior corrosion resistance and its strength characteristics make it an optimal material for uses that need maximum toughness. It is not normally used in art or architecture unless it is incorporated into protective barriers.

STAINLESS STEEL ALLOY S31803

Duplex Alloy

UNS S31803 UNS S32205	ASTM A182 ASTM A240 ASTM A276 ASTM A479
2205 EN 1.4462 SUS 329J3L	

Alloying Constituents Added to Iron (wt. percentage)

C	Cr	Ni	N	Mo	Mn	Si	S
0.03% max	22–23%	4.5–6.5%	0.14–0.20%	3.0–3.5%	2.0% max	1.0% max	0.02 max

	Tensile strength		Yield strength		Elongation % (2 in.–50 mm)	Rockwell hardness
S31803	90 KSI	621 MPa	65 KSI	448 MPa	25	—
S32205	95 KSI	655 MPa	65 KSI	448 MPa	25	—

Available Forms

Sheet

Plate

Bar

Tube

Pipe

Alloy S31803, also known as S32205, is often considered for use in art and architecture exposed to chloride conditions. This alloy has exceptional pitting resistance. It is harder to form, and shape and work hardens rapidly. Of the corrosion resistant alloys, in particular those resistant to chloride ion environments, S32205 is often specified. It is harder to work than the austenitic alloys and allowances for springback and for bend radius should be understood when fabricating with S32205 and other duplex alloys, as they are more difficult to form than the common stainless steels used in art and architecture.

ARCHITECTURAL CAST ALLOYS

Stainless steel castings are not as common in art and architecture as the wrought forms of the metal. They are difficult to finish. All post-finishes on casting require some level of mechanical surfacing. For casting alloys of stainless steel, the UNS system uses J as the lead letter for cast steels followed by the number 9 for the alloys of stainless steel castings.

There are a number of cast stainless alloys similar to the wrought alloys; however, there are subtle differences from their wrought cousins in terms of chemistry. It is advisable to specify the cast alloy rather than the equivalent wrought alloy. The equivalent wrought alloy is included in the description of a few common cast alloys for reference only. Below this will be listed the equivalent number designation established by the Alloy Casting Institute (ACI). In this section only alloys of stainless steel that most correspond to the common austenitic alloys used in art and architecture are included. There are many cast stainless steel alloys available and in common use in selected industries. Different alloys were developed for specific industrial applications as understanding of the value of stainless steel increased.

All the cast alloys listed below can be heat treated and often are. Cast stainless steel differs from wrought stainless steel mainly in the microstructure of the grains. In wrought alloy forms, the grains can be cold worked to develop strength and refinement. In castings, the microstructure is nonuniform and nonhomogenous due to the way the molten metal cools and solidifies. In wrought forms you can modify this and make the microstructure more homogeneous by processes of cold working and annealing. This is not possible with castings.

All of the cast alloys listed can be heat treated to modify their as-cast microstructure somewhat, to improve corrosion resistance and strength characteristics. For these alloys the heat treatments are solution heat treatment processes, which bring the castings up to temperatures of 1040–1150°C, depending on the alloy. They are then rapidly cooled by quenching in water. It is important to work closely with a foundry to establish the precise needs and properties desired.

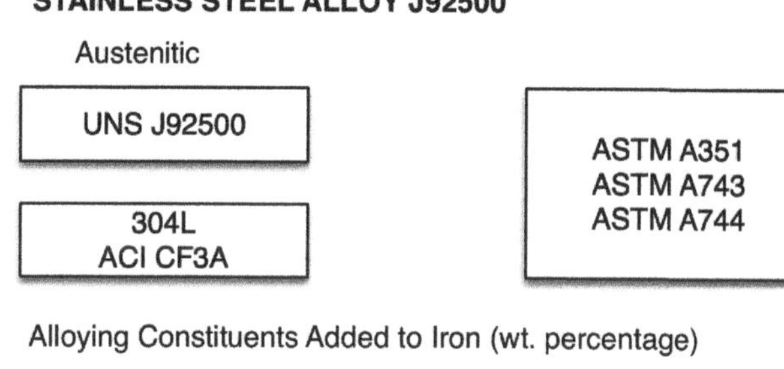

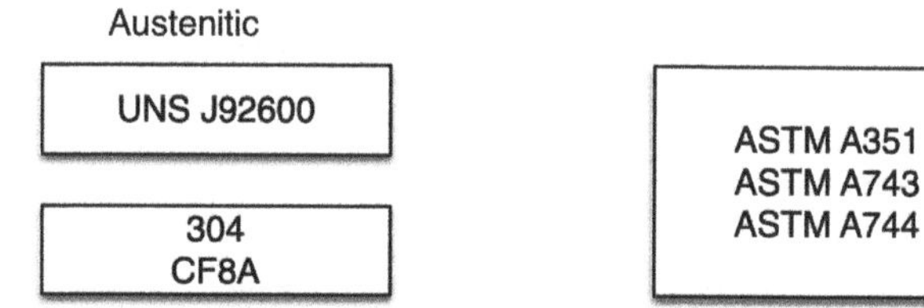

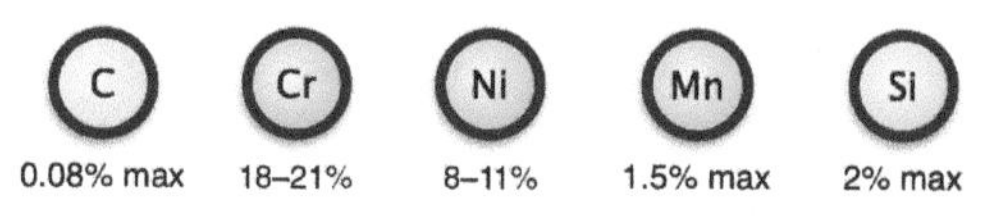

	Tensile strength		Yield strength		Elongation % (2 in.–50 mm)	Brinell hardness
J92500	77 KSI	531 MPa	36 KSI	248 MPa	60	140
J92600	77 KSI	531 MPa	37 KSI	255 MPa	55	140

These alloys are similar to the wrought alloys S30400 and S30403. They are common cast alloys of stainless steel. They are usually solution heat treated to achieve ideal characteristics.

STAINLESS STEEL ALLOY J92701

Austenitic

UNS J92701

303Se
CF16F

ASTM A743

Alloying Constituents Added to Iron (wt. percentage)

0.16% max

18–21%

9–12%

1.5% max

2% max

0.17% max

0.04% min.

0.2–0.35%

	Tensile strength		Yield strength		Elongation % (2 in.–50 mm)	Brinell hardness
J92701	77 KSI	531 MPa	40 KSI	276 MPa	52	150

Alloy J92701 is lower strength but has good machinability. Extensive post-machining of the casting is best suited for this alloy of cast stainless steel. Usually there is 1.5% molybdenum along with 0.2–0.35% selenium.

STAINLESS STEEL ALLOY J92800

Austenitic

UNS J92800

316L
CF3MA

ASTM A351
ASTM A743
ASTM A744

Alloying Constituents Added to Iron (wt. percentage)

0.03% max

17–21%

9–13%

1.5% max

1.5% max

2–3%

STAINLESS STEEL ALLOY J92900

Austenitic

UNS J92900

316
CF8M

ASTM A351
ASTM A743
ASTM A744

Alloying Constituents Added to Iron (wt. percentage)

0.08% max

18–21%

9–12%

1.5% max

1.5% max

2–3%

	Tensile strength		Yield strength		Elongation % (2 in.–50 mm)	Brinell hardness
J92800	80 KSI	552 MPa	40 KSI	276 MPa	55	150
J92900	80 KSI	552 MPa	42 KSI	290 MPa	50	170

These cast alloys are similar to the wrought alloys S31600 and S31603. They have good strength and excellent corrosion resistance. The low carbon grade, J92800, is more suitable for welding.

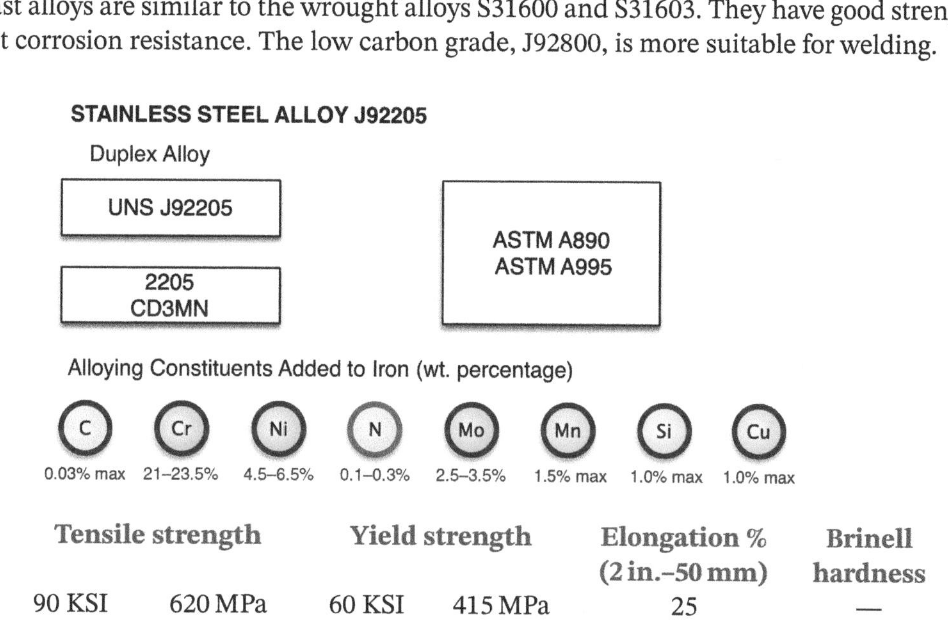

	Tensile strength		Yield strength		Elongation % (2 in.–50 mm)	Brinell hardness
J92205	90 KSI	620 MPa	60 KSI	415 MPa	25	—

This duplex alloy is considerably stronger than other versions of stainless steel cast alloys. It has excellent corrosion resistance as well.

There are many more alloys of stainless steel not listed here, both wrought forms and cast forms. Those listed, however, are the alloys used in art and architecture—or that lie just off the edges of common use. These alloys are the North American versions. There are European and Asian versions that may differ slightly in constituents

CHAPTER 3

Surface Finishes

The aim of art is to represent not the outward appearance of things, but their inward significance.

Aristotle

METALLIC LUSTER

Perhaps the single greatest attribute of a stainless steel surface is its consistent and predictable interplay with light. Stainless steel reflects approximately 60% of a light wave. But more importantly, the luster generated by the absorption and reemission of a light wave intensifies the appearance of stainless steel, providing it with an exceptional aura few other materials can match. This characteristic of absorption and reemission is one all metals possess. Metals conduct electricity and heat more readily than other materials because they have valence electrons: electrons in the outer shell of the atom that delocalize from their respective atoms and are shared. This has been likened to a "sea of charges" around the atoms that make up a metal (Figure 3.1).

When light falls on a metal surface it is intensely absorbed only by the surface atoms. Light penetrates only a tiny distance: less than a wavelength. This intense absorption of the electromagnetic wave of light on the surface causes a pulse of alternating current, which then excites this "sea" of charged particles and reemits the light. This is luster, the intense reflection from a polished metal surface. The smoother the surface, the more the reflection. If the surface is coarse, a diffuse reflection occurs. With minimal care, this luster will remain part of the surface of stainless steel for many decades to come; removal of dirt and grime or oxides that develop from salt reactions will again make the luster apparent.

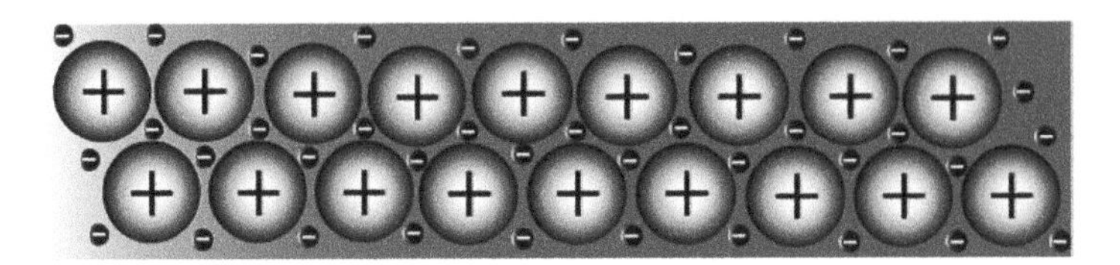

FIGURE 3.1 Sea of shared electrons around metal atoms.

The single greatest attribute of a stainless steel surface is its interplay with light. Stainless steel reflects approximately 60% of a light wave. More importantly, the luster generated by the absorption and reemission of the light wave intensifies the appearance.

FINISH ON THE SURFACE

One of the most attractive things about using stainless steel in art and architecture is the plethora of options. No other metal offers the vastness and variety of options available for stainless steel. The vast array of finishes, textures, and colors available can be exponentially expanded when used in combination.

For the purpose of definition, a finish is a mechanical or chemical treatment on the surface that imparts a particular interaction behavior with light. Mill finishes are applied to the surface of the wrought metal form by the Mill producer as the metal is cold worked and reduced, then annealed and pickled. For all intents and purposes the mill finish is the base for all subsequent finishes. The mill finish can and often is left untouched. It is a stable surface created early in the life of the metal. It is unique to the time and place of production and cannot be matched in subsequent castings or mill runs. Mechanical finishes are the next common finishes on stainless steel. These are applied to the surface by sanding, glass bead blasting, and buffing at a secondary processing facility that receives the metal from the Mill. These cover a wide range of finish possibilities, from linear scratch lines produced rapidly to glass bead produced in a controlled chamber to highly polished buffed mirror finishes. Mechanical finishes also include an array of custom finishes produced mechanically under computer control.

In the last half of the twentieth century, techniques of adding color to the stainless steel were commercially established. These techniques involved no pigments, dyes, or paints, but instead capitalized on the phenomena of light interference induced by thickening the natural, transparent, chromium oxide film. Interference coloring, as it is sometimes referred to, can be applied to any of the stainless steel alloys and to any finish, thus further expanding the possibilities for design with stainless steels. All of these finishes and all of these techniques have interaction with light as the primary feature of interest.

NATURAL COLOR

The natural color of stainless steel is derived from the interaction of light with the chromium and iron in the surface. Essentially, when two or more metals with the same structure are mixed together so that they form a solid solution, the resulting color will be something intermediate between the color of the metals. Color theory for metals has much to do with the absorption of photons by the outer electrons, or d-shell, of the particular metal. The electrons in the d-shell jump to the s-shell. It is not fully understood and gets deep into quantum physics, but the reason that stainless steel has a silver sheen is due to the way that the chromium atoms on the surface absorb and reflect light. The reflectance of steel is similar to that of stainless steel, although lower across all visible wavelengths. This gives polished steel a gray tone darker than stainless steel, because it is the chromium that provides stainless steel's color and luster. Chromium stabilizes the surface as it develops the clear chromium oxide protective layer. Chromium has a high reflectance that is slightly above that of stainless steel, so the net effect is what is shown in Figure 3.2. Nickel and molybdenum have a very slight dulling effect on the luster. It is the chromium content that really provides the deep metallic luster we associate with stainless steel.

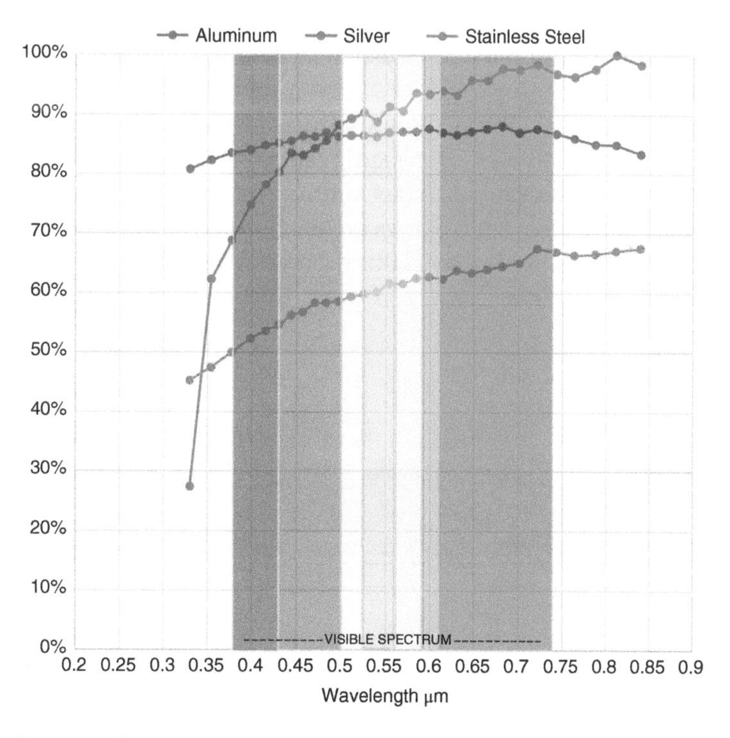

FIGURE 3.2 The reflectance of stainless steel, aluminum, and silver.

Stainless steel's natural silver luster can be enhanced by expanding the chromium oxide layer through a chemical interaction with the surface. There are several methods in use and all interact with the chromium oxide layer. International Nickel Corporation (Inco) developed a process that produces color through the phenomenon of light interference. It patented the process in the early 1970s. The process involves treating the surface of the stainless steel in a heated bath of a proprietary mixture that involves chromic acid. This results in the development of a thickened layer of clear chromium oxide. This layer interferes with the light wave as it strikes the thickened oxide layer. A portion of the light wave passes through the clear oxide, while a portion of the wave reflects off the surface of the oxide. As these two offset waves interact as they reflect back to the viewer, a phenomenon of light interference occurs. A portion of the wavelength is canceled out while another portion may be enhanced. The viewer sees only the portion that has not been canceled out through what is known as "destructive interference."

Commercial enterprises, expert in the process controls, are able to home in on a specific color and lock it by stopping the process at a particular film thickness. These interference colors cannot be just any color, however. Since the process involves interference with a light wave, no pigmentation or dye is introduced to produce the color. This is both a benefit and a constraint. You are limited to the tight range of colors in the light spectrum: thus, you cannot choose a color that combines the spectral colors as you can when mixing pigments. Color produced in this manner – through the interplay of light and the chromium oxide layer – will never fade. There is nothing to decay from exposure to ultraviolet rays (Figure 3.3).[1]

There are additional industrial methods in use to apply color to stainless steel. Paint is rarely used, except where corrosion of the metal is a significant concern. Stainless can be painted by using special primers that adhere to the stainless steel surface. In interior applications and for some artwork, transparent and semitransparent coatings are used to take advantage of the reflective quality of stainless steel. Jeff Koons uses highly specialized processes to apply semitransparent coatings to his art.

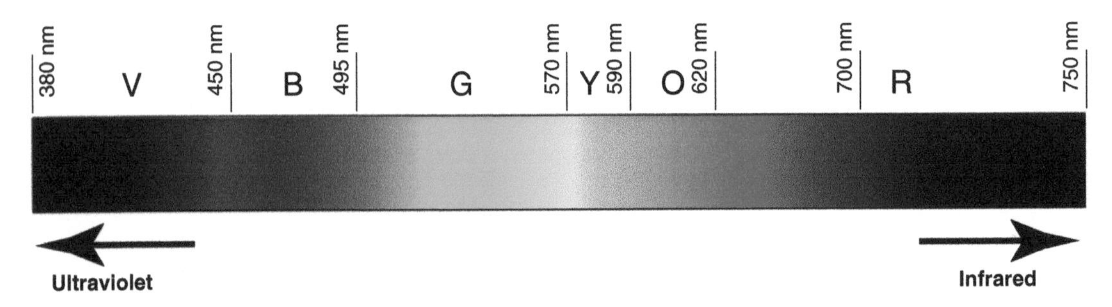

FIGURE 3.3 Light spectrum and wavelength corresponding to color.

[1]There is some evidence that the color of stainless steel has deteriorated in desert regions as a result of surface dehydration. There is also evidence of some of the thicker films of chromium oxide that produce green and red colors showing premature deterioration in heavy salted regions.

Another commercial process used to add color to stainless steel is physical vapor deposition, or PVD. PVD is a term that covers several processes. In each of these processes a very thin layer of another metal or metallic compound is applied to the surface using a technique that vaporizes a substance and then causes that substance to condense on the stainless steel surface. The colors are induced from a combination of thin-film interference and the color within the substance. Gold, for example, is the color imparted by titanium nitride (TiN) PVD (Figure 3.4). These processes are

FIGURE 3.4 PVD gold (titanium nitride) coating over Angel Hair stainless steel; designed by Leong Leong Architects.

"line of sight" processes, in which the thin coating is applied to exposed areas. PVD is not a dipping technique and sheets of stainless steel coated by physical deposition are coated on one surface only.

There are other chemical applications that can be applied to the stainless steel surface to produce blue and black hues, although they are not extensively used in art and architecture because of inconsistent results. These techniques, often referred to as bluing or dark chroming, involve dichromate salt baths to produce thickened oxides of chromium. There are cold applications as well that put a darkened, mottled stain on the surface (Figure 3.5).

Since the metal was first conceived and successfully used on the Chrysler Building roof and spire in New York in the late 1920s, the expectations have been that once installed, little needs to be done. There are many examples where little other than natural rainfall have kept the surfaces looking quite pristine; on closer examination, though, you will see the beginnings of stains or the buildup of grime. The occasional cleaning, similar to what is needed on a glass façade or a painted metal veneer, will keep the stainless steel surface looking impeccable. On interior surfaces and exterior surfaces where hand prints might leave their oils and residues, cleaning is required to keep the metal luster as vibrant as when first installed. Fingerprint oils and other mild organic residues will not hurt stainless steel; they only add a layer of oil that interferes with reflected light, making them

FIGURE 3.5 Residential building at the High Line in New York City, designed by Zaha Hadid Architects.

more apparent. A protocol of regular washing of the stainless steel surface does not have to be an expensive or elaborate process. Stainless steel is easy to clean and will remain so with little effort.

ONE SIDE PRIME FOR FLAT PRODUCTS

It is important to note that the surface quality of sheet and plate material is established on one surface: the prime surface. One can specify a two-sided finish sheet or plate, but at a significant increase in cost. Often the back side is of lower quality due to the way the Mill produces the initial finish. The lower roll used at the Mill is not as refined as the upper surface roll. The opposing surface is not considered within a standard of quality. There are no recognized standards for the back surface of a metal sheet or plate, and typically this has little relevance since most sheet and plate are used as cladding for structural or surface components or are folded around to produce tubular forms with a hidden interior.

All industry specifications describe and evaluate the finish of stainless steel on the prime surface. The reverse surface can have mars, slight scratches,[2] and other surface imperfections. The reverse side should be clean and free of scale or polishing compounds, but it is not much more refined than a mill finish. Fabrication facilities working with flat sheet, coil, and plate work with the prime side facing upward, visible for inspection. This can lead to marring or mild abrasion on the reverse side, but normally imperfections are not visible and do not manifest to the face side.

In most applications this non-face surface is hidden from view, but the occasional perforated wall panel may subject this surface to closer scrutiny. In these instances the sheet or plate can receive a polish on this side. The non-face side will need to be protected during fabrication in a fashion similar to the prime, face side. The cost will be higher due to the additional polishing and handling. The back surface will not match the face surface.

Some directional satin finishes will put a light polish pass on the reverse surface. This is called a "back pass." This back-side finishing is needed to stress relieve the metal sheet and to balance out the stress created on the face side due to the polishing. It lacks quality and consistency. This is also necessary for glass bead blast finishes. Blasting the reverse side balances out the stresses on the face side. If not done, the sheet or plate will warp and bow from the differential tensioning on one surface versus the other.

One of the most attractive things about using stainless steel in art and architecture is the plethora of options. No other metal offers the vastness and variety of options that are available for stainless steel. Many of the surface finishes can be used in combination with other finishes to exponentially increase the options available.

[2]The caution here is that slight scratches and even mars can be apparent on the face side of thin sheet metal under certain lighting conditions.

THE NEED FOR A CLEAN SURFACE

When performed correctly, for the most part mechanical and chemical finishing have little intrinsic effect on the corrosion resistance of the metal. However, some of the finishes do increase the maintenance necessary to keep the metal clean and free of potentially corroding materials. The coarser the finish, the more it will hold pollutants to the surface (Table 3.1 shows the relative coarseness of various finishes). Coarse surfaces are also harder to clean because it takes more energy to break pollutants free of the microscopic grooves and undercuts. A coarse surface can also create capillary condensation and hold water in its small grooves. When these finishes are applied to the metal's surface, they must be applied with clean brushes, belts, and blast media. They must be applied to very clean surfaces to eliminate the potential for embedding contaminants into the finish. Care must be taken to eliminate microscopic undercutting, which allows dirt and other pollutants to bind to the surface. For the fabricator of stainless steel, having a very clean surface prior to mechanical finishing operations is critical.

The uninitiated often feel that mechanical finishing operations such as glass bead blasting or abrasive brushing will remove surface contaminants, making a degreasing and cleaning step unnecessary. In reality, the finishing operation only conceals contaminants and can make conditions that affect performance in the future worse. Finishing operations do not remove a significant amount of metal and alter surface reflectivity. Under a microscope, contamination is often seen to be still present. Fingerprints, for example, will not be removed with glass bead blasting; instead,

TABLE 3.1 Relative comparison of surface roughness of various finishes.

Finish	Relative coarseness (1 = low; 5 = high)
Highly reflective: 2BA, 7, 8, 9	1
2B	1
Rolled finishes: linen, others	2
2D	2
Hairline	2
Angel Hair	3
Glass bead	3
Stainless steel shot blast	4
No. 4	4
No. 3	5
Grind swirl finish	5

they can be "locked" onto the surface where they can appear as ghosting. Electropolishing is needed to remove them.

Surfaces newly finished with glass bead blasting and chemically etched surfaces absorb oils, whether from processing or handling. Removal of the oils and cleaning of the surfaces can be major tasks. When the outer chromium oxide is altered or removed chemically, the porous surface will want to grab and hold substances. These finishes, if exposed to the atmosphere for a few days, will become resistant to fingerprinting. The surface attracts moisture from the air and this newly hydrated surface is no longer attracted to oils from fingerprinting. For interior applications this is generally not the case. Interior applications tend to remain subject to fingerprinting.

ANTI-FINGERPRINT COATINGS

There are several anti-fingerprint coatings available for stainless steel surfaces. These thin nano-coatings seal the surface from light oils and fingerprints. They perform exceptionally well as long as they are not scratched or marred deeply enough to remove the finish. They are very durable nonetheless and will hold up both in interior and exterior applications where light abrasion from human interaction is prevalent.

These coatings effectively seal the microscopic pores that exist in a stainless steel surface. They are hard coatings. Most remarkable is their even consistency and invisibility to the eye. They are truly nano-coatings, measuring from 5 to 10 μm in thickness. They are hydrophilic films that seal the surface from oils, moisture, and other grease.

They will not prevent attack from chloride salts or deicing salts. Anti-fingerprint coatings will alter the sheen and tone of the stainless steel surface slightly, but consistently, across the surface. These coatings are applied to stainless steel surfaces that have a light to moderate texture. These coatings are not suitable for highly reflective surfaces. They have a slight optical effect and can make mirror surfaces cloudy. They cannot be used on interference colored stainless steels or PVD colored stainless steels.

These coatings are very thin, silicate-based films that are applied wet and then activated by exposure to ultraviolet light or thermal curing. They are usually applied in a plant, as currently there are limits to in situ coating. These coatings have good thermal expansion characteristics, good fade resistance, and can be formed. Some micro cracking will be present, but this cracking is invisible to the naked eye. The coatings themselves are dielectric in nature, but they do not perform as dielectric coating separators for dissimilar metals in contact with one another.

POST-FINISHING PROTECTION

Post-finishing operations would benefit by keeping the metal in a clean, dry place for a period of time, to allow the chromium oxide layer to develop. Quickly coating the surface with an impervious layer of plastic film can actually hamper the development of the oxide. Immediately coating stainless steel that has received a bead blast finish can alter the oxide development to the point that the appearance of the stainless steel will differ from other sheets that have been bead blasted and allowed to remain exposed to the atmosphere for a few hours. Instances where a sheet has been left exposed overnight while others were protected have created a permanent difference in appearance. Wrapping newly finished sheets with plastic but leaving small bubbles of air between the metal and plastic will create spots where the oxidation has occurred in the bubbles at a different rate than in the balance of the sheet.

A mild electropolishing operation post–mechanical finishing can render a passive surface. Electropolishing removes undercuts and deleterious surface imperfections. It will also brighten the surface, giving the metal a super clean appearance.

STAINLESS STEEL FINISHES

Finishes on stainless steels fall into either one or a combination of the following categories (Table 3.2):

- Mill finishes
- Mechanical finishes
- Chemical finishes
- Plastic deformation surface finishes

Each of these finishes has its own particular nuances, but they all start at the Mill. The Mill is the place where scrap is melted and the metal elements of the alloy are added. The real metallurgy occurs at the Mill. The Mill and not the mine where ore is removed from the earth is the beginning. The Mill today is a high-technology, modern facility that takes recycled scrap and refined ore and formulates a molten mixture of various elements, pumps oxygen, argon, and other gases to remove contaminates, and arrives at a massive casting: a block of alloy steel. This block is fresh from the

TABLE 3.2 Various finishes available.

Mill finishes	Mechanical finishes	Chemical finishes	Plastic deformation
No. 1 HRAP	No. 3	Interference coloring	Embossing
No. 2D	No. 4	PVD	Custom Embossing
No. 2B	No. 7	Bluing	Coining
No. 2BA	No. 8	Heat tinting	Water
	Super Mirror		Custom
	Custom		

furnaces and nothing matches it; like a fingerprint, this block of steel alloy has never been produced before and it will never be produced again. Alloys of steel, cast in massive furnaces, are of their particular time and space. They are predictable within very narrow criteria, but like the trees in a forest, no two are alike. The Mill is the birthplace of stainless steel.

MILL FINISHES

Mill finishes are the finishes provided directly from the original source of the metal on stainless steel sheet, plate, pipe, tube, extrusion, and casting. In reality they are a combination of basic mechanical and chemical treatments to stainless steel that impart the initial genesis of the metal's surface (Table 3.3). Critical controls at the Mill will lead to desirable results in the finish of an art or architectural feature made from stainless steel.

Different levels of surface character are achieved depending on the form produced by the Mill and the standards of quality the Mill adheres to. The steel Mill wants to sell tonnage. It wants to produce a metal to tolerance standards and then move it out. It is not interested in fussing about with the peculiarities an artist might have in mind for the surface of the material. These are massive facilities that are geared to work in bulk. The work they do is like growing a forest in a nursery: organized, efficient, and free of contaminating substances.

There is a fascinating story about the great architect Louis Kahn – perhaps one of the most influential architects who ever lived – when he was designing the Yale Center for British Art. He wanted a special surface for the façade and wanted to know his materials of design. He went to the Allegheny Ludlum stainless steel mill in Pennsylvania. He saw some stainless steel at the Mill, still in the hot rolled state, and wanted it for his façade. The metal was then set aside and used on the surface.

TABLE 3.3 Type of finishing on stainless steel.

Type of finishing	Specific form	Light reflective behavior
Mechanical	Grinding	Material removal – roughening
	Dry polishing	Polarized scattering
	Wet polishing	Fine scattering
	Buffing	Mirror reflectance
	Non-directional (Angel Hair, glass bead)	Light scattering
Chemical	Electropolish	Increase reflectivity
	Acid etching	Matte, scattering
	Chromic acid dipping	Color
	PVD	Color

FIGURE 3.6 The Yale Center for British Art, designed by Louis Kahn.

In a sense, that metal is one of a kind: a metal produced at a particular Mill at a particular time in a hot rolled condition. Some have said it was to be rejected and recast because the quality was not consistent across the plates. Just ordering hot rolled and pickled stainless steel in an attempt to match that metal is not possible. It was produced at a particular time and under the conditions of that time. It has an interesting light and dark character running the length of the sheets that give the surface a remarkable quality (Figure 3.6).

Hot Rolled, Annealed, and Pickled (HRAP)

For the wrought materials – plate, sheet, and rods – the steel mill first casts a large billet or bloom of a particular alloy of stainless steel. This casting is thick: usually from 125 to 400 mm in thickness.

Once cast, the slab undergoes a high-speed grinding process to improve the surface of the slab and to remove oxides and scale prior to rolling. The first rolling process encountered by the slab is hot planetary rolling, which involves passing this red-hot slab of stainless steel under pressure applied by planetary rolls. This process reduces the slab thickness to approximately 6 mm. The steel is then rolled into a coil of a thick, undulating ribbon. The metal is not very flat, and the surface is rough and dark. The coil is placed into an annealing oven to remove the stresses that have built up in the metal. Annealing also reduces carbides that have precipitated during cooling by taking them back into solution. Annealing is a controlled high-temperature heating of the metal. For the austenitic stainless steels, the annealing temperature is between 425 and 815°C (800 and 1500°F). This temperature softens the metal for subsequent cold rolling processes. After annealing, the coil is pickled to remove scale. Scale is a hard coating of oxide that forms on the surface of the steel during annealing and subsequent cooling. Pickling involves strong oxidizing agents, such as hydrofluoric acid and nitric acid, to remove this scale prior to cold rolling. The edge of the ribbon of metal is often trimmed and recycled. This initial edge is uneven and wavy. Trimming refines the coil dimension. The initial cold rolling operation passes the ribbon of metal though pressure rolls and further thins the material to approximately 4.5 mm. The reduced stainless steel ribbon is sanded on both surfaces to eliminate surface inclusions that have developed from the hot rolling process. Hot rolled stainless steel often contains small pits across the surface, and flatness is not to the levels obtained in cold rolling. After sanding and cold rolling, the coils of metal are again annealed and pickled to arrive at a more finished surface and to remove the work hardening that has occurred in this initial cold rolling.

Further cold rolling is performed on these annealed coils, reducing the thickness to approximately 3 mm. Stresses develop in the sheet during this cold rolling, cold reduction process, requiring an additional annealing and pickling step. For architectural and art projects, this step in the process is of critical importance in achieving good quality surfaces. The annealing stage at this point in the operation will impart the final color tone to the stainless steel. It is recommended that all coils be annealed and pickled in one single batch, which will achieve a final color uniformity in the initial stainless steel mill surface by insuring that all the metal has been subjected to the same conditions of time, temperature, and annealing atmosphere. For large projects it is simply not possible to anneal all the necessary coils in the same oven. Additionally, minor fluctuations in the alloy mix (still within industry standards) can lead to appearance variations in certain lighting conditions. The mill surfaces in particular are less refined than the finished surfaces that undergo further polishing and abrading.

This initial surface is sometimes referred to as the "hot rolled, annealed, and pickled" (or "HRAP") surface. The HRAP surface is coarser and lacks smoothness. It possesses the interesting character of a thick steel surface, lacking the refining and polishing processes but retaining the toughness steel is known for. This finish is also referred to as a No. 1 finish.

After the final annealing and pickling step, the coils are passed through a set of cold reducing rolls. This step further thins the metal and introduces strength by means of cold working the metal. This step also imparts the final surface finish provided by the Mill. The stainless steel coils are degreased and, if desired, are placed in a bright annealing oven to achieve a further refined mirror-like surface that is referred to as "BA," or "bright annealed."

FIGURE 3.7 Sheet leveling equipment.
Source: Photo courtesy of Rimex Metals.

The stainless steel coil is then passed through a sheet leveler to remove stresses induced by the coiling process, protected with a plastic film, and sheared down to sheet size, where they are stacked on a pallet (Figure 3.7).

Mill Sheet Finishes

The mill sheet finishes are 2D, 2B, and 2BA. These finishes are available on stainless steel sheet material with a maximum thickness of 4.75 mm. These finishes are produced at the Mill source and are provided as is for use in art and architectural projects or as a base for further refinement.

2D Finish

For thin sheet, No. 2D finish is the least refined surface available for sheet material. 2D surface finish is available in sheet, coil, and strip. The 2D surface is dull and characterized by a dark steel-like appearance. It is sometimes used on roof surfacing materials or other cladding surfaces where light reflectivity and the associated glare are required to be minimized. The 2D surface is not refined and can possess light streaking along the length of the surface. It is produced by passing thin sheet material through unpolished cold rolls. On thin sheet metal that is less than 1.5 mm thick, the surface often possesses a slickness or gloss. It is important in artistic and architectural applications for the Mill supplier to discontinue the operation of ink marking the surface with alloy or specification marks. Mills often mark the alloy and specification by printing identification criteria on the metal surface. Unless you want your roof to be marked with printed lettering, the Mill should be made aware of the intended use. 2D finish is one of the more economical finishes and is used as a base finish for pattern finishes and other mechanically applied finishes where the desire is for a low reflective appearance. This mill finish is not repairable if damaged. No chemical process or mechanical finish will restore a damaged 2D surface. The only alternatives are to live with the surface damage or to give the surface a different finish by mechanically abrading it.

This finish does not receive color well through light interference coloring techniques or PVD coloring techniques. It will color, but mottling and streaking are to be expected (Figure 3.8).

2B Finish

This mill finish is produced by passing the 2D surface coil between polished rolls on its final pass. This finish is characterized by a surface more refined and glossy than 2D. 2B finish is a base

FIGURE 3.8 No. 2D stainless steel roof on Fresno City Hall, designed by Arthur Erickson.

finish for many mechanically applied finishes. Occasionally the 2B finish is used on an artistic or architectural surfacing material without additional mechanical treatments. It is used as a finish in commercial kitchens, not for its aesthetic appearance, but for surfaces that are subject to mars and scratches from pans and utensils being passed over them. It is used on countertops and tables where these small scratches imparted from objects are accepted as part of their appearance. The surface is subject to a phenomenon known as "coining"; that is, when a metal edge is passed over the surface, a mark is imparted onto the surface. Coins will do this to stainless steel trays, but when one sheet edge is passed over the surface of stainless steel a coining streak is left. This coining mark or mar is not reversible. It cannot be removed except by mechanically abrading or polishing the surface.

Like 2D mill finish, 2B mill finish is not repairable if damaged. Also, surfaces with a 2B finish cannot be matched to each other through polishing or chemical treatments because 2B is a naturally developing finish imparted to stainless steel at the Mill. It is created on the surface as the stainless steel rolls between semifinished rolls. Once scratched, marred, dented, or welded, the only options are to change the finish or live with the damaged surface.

2BA Finish

This is the most refined of the mill finishes. It is characterized by a mirror-like reflectivity. The 2BA, or bright annealed, surface is very smooth and generally free of surface markings or streaks. The finish is produced by cold rolling a 2B finish on highly polished rolls, then annealing the coil of metal in a controlled atmosphere, usually of hydrogen or nitrogen. This processing keeps the surface bright and smooth.

2BA is used as a base for many mechanical finishes and chemical finishes, including mirror polishing, satin polishing, glass bead blasting, and Angel Hair polishing.

It is used in art and architecture when a clean refined surface is required. 2BA finishes, like the other mill finishes, are not repairable. It acquires its mirror-like quality from both the final pass under pressure through the polished roll and its subsequent annealing in a controlled atmosphere to give the surface a unique, near natural character. Scratches, mars, and dents will alter this finish permanently. As with 2B finish, the only options are to live with the damage or change the finish.

Variations

In Europe they sometimes refer to the 2BA finish as 2R. Additionally, some Mills make a further distinction by calling a finish BA only if it undergoes heat treatment in a controlled atmosphere furnace. These Mills refer to 2BA as a finish that has been heat treated in a controlled atmosphere furnace and then cold rolled through highly polished rolls.

Some 2BA finishes lack clarity. They are reflective, but the reflection is cloudy. Obtain representative samples from the supplier to insure that the Mill source is capable of providing the surface required (Figure 3.9).

FIGURE 3.9 Examples of No. 2D, No. 2B, and No. 2BA stainless steel.

MECHANICAL FINISHES

Mechanical finishes are those produced by a secondary finishing facility. These facilities take the mill finish stainless and apply a second, more refined surface to the prime side of the sheet or plate. Considered repairable, there are procedures in the event of scratches or mars for most of these finishes. Repairs are extremely difficult to execute in situ. On large flat areas, repairs can cause more problems than the initial damage. Thus, proceed with extreme caution when attempting repairs to damaged mechanical surfaces. Many mechanical finishes can camouflage the appearance of minor surface imperfections from most viewing angles.

There are four categories of mechanical finishes:

1. Directional
2. Non-directional
3. Custom
4. Mirror

DIRECTIONAL FINISHES

Directional finishes are distinguished by parallel, linear scratches that run the length of the sheet. The scratches are of regular and consistent depth. Light reflecting off of these surfaces is stronger in one direction than the other direction due to the directional surface ridges that define these finishes. Three of the most common are described in Table 3.4 and in a later section, and shown in Figure 3.10.

There are other directional finishes, such as No. 6, that have fallen out of use. The No. 6 finish was produced by Tampico brushing the surface of stainless steel. Tampico is a stiff fibrous material produced from *Agave rigida*, a yucca-like plant. Today, the nomenclature "No. 6" is sometimes used incorrectly to label a finer version of a directional No. 4.

Directional finishes are produced in coil, sheet, and post-applications to shapes and forms. These finishes can be applied to tube, pipe, bar, and structural shapes. They can be applied over any of the mill surfaces, but are usually applied over 2B and 2BA. Due to its coarse surface, the No. 3 finish is typically limited to interior applications where welding and finish blending may be required.

The reflectivity of directional finishes is polarized; that is, they reflect more strongly in one direction than another. They readily fingerprint, and they will hold deicing salts and other contaminants. Thus, regular cleaning is necessary.

The directional finishes are applied using belts impregnated with abrasives of different grit sizes. On stainless steel there are typically two types of grit used: silicon carbide and aluminum oxide.

TABLE 3.4 Common directional finishes.

Finish	Description	Typical Ra
No. 3	Coarse	Max. 40
No. 4	Fine	Max. 25
Hairline	Very fine	Varies

FIGURE 3.10 Examples of No. 3, No. 4, and Hairline directional finishes.

These substances are harder than the surface of stainless steel and produce a signature scratch profile. Note that aluminum oxide and silicon carbide will have distinctly different reflectivity even when a similar grit is used. Figure 3.11 gives a representation of 120-grit aluminum oxide versus 120-grit silicon carbide applied to stainless steel. The aluminum oxide finish appears lighter in color due partially to the greater light scattering of the aluminum oxide finish. When polishing with an aluminum oxide abrasive, you must continue with that same abrasive. You cannot match it with a

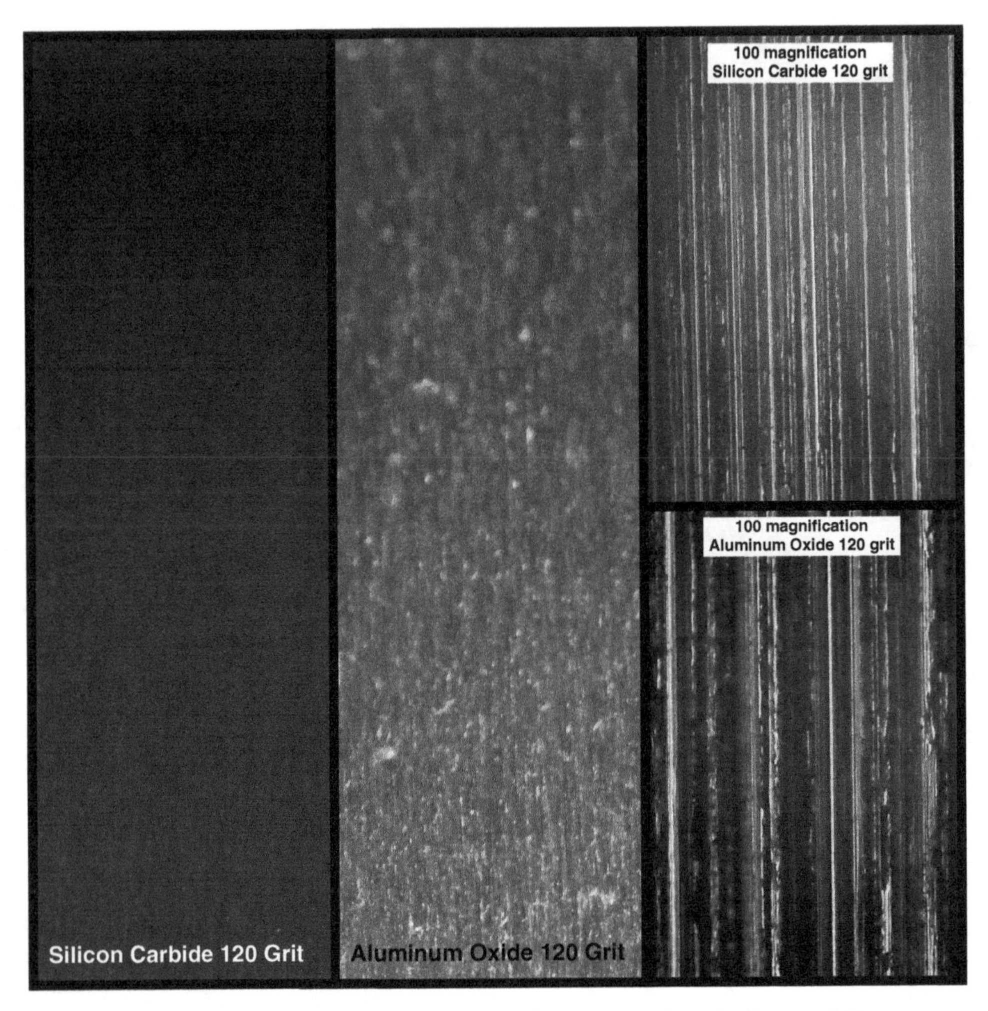

FIGURE 3.11 Comparison of a 120-grit finish applied via aluminum oxide and silicon carbide.

silicon carbide abrasive. Aluminum oxide results in a coarser microscopic texture than silicon carbide, and it tends to create more undercuts and rips. Post-electropolishing the surface will remove these, smooth out the surface, and at the same time make it more reflective (Table 3.5).

It is recommended to get representative samples of these finishes and discuss the quality processes to insure a tight range. Companies often resort to using roughness parameters as a guide to define their processes. These should be used as a guide rather than an approval or disapproval. Small variations can be meaningless to the end result.

There are distinct differences in the directional finishes produced by aluminum oxide and silicon carbide. The aluminum oxide has a deeper cut. It is less reflective than the comparable silicon carbide. The gloss reading for a 120-grit silicon carbide finish may be in the range of 68–78 while

TABLE 3.5 Comparison of roughness and gloss readings.

Finish type	Roughness reading Ra	Gloss (60°) perpendicular to grain	Gloss (60°) parallel to grain
120-grit aluminum oxide	9.5	13	96
120-grit silicon carbide	6.5	76	284

the gloss readings for 120-grit aluminum oxide finish can read as low as 15–20. This is a significant drop in reflectivity.

Aluminum oxide leaves small skipping parallel lines while silicon carbide leaves long lines. There is somewhat more tearing of the surface with aluminum oxide versus silicon carbide, and when passing a microfiber cloth over the surface of an aluminum oxide finish the cloth will snag on the minute tears. The aluminum oxide finish will hide dirt and streaks better than the silicon carbide finish because of the roughness of the surface.

There are no clearly defined standards for polishing, buffing, and brush polishing stainless steel surfaces. Grit and Ra (roughness average) are often used as representative values specified as targets. Grit represents the abrasive size used; however, the same grit can produce a variety of surface appearances and reflectivity. The finish is highly dependent on machine type and speed, pressure, and operator skill. Using the same grit you can achieve a matte finish or a bright reflective finish. It is recommended to match up samples of different grits and polish types and view them from different angles to validate the reflective appearance.

Terms used with respect to finishing stainless steel are grinding, polishing, buffing, and brushing. Each represents a different process and finish, and each can be subdivided into different levels (Table 3.6).

Typical roughness parameters used are:

Ra	Average depth of the scratch
Rmax	Maximum depth of the scratch
Rz	Uniformity of scratches
RMS	European measurement; similar to Ra

Surface roughness is a component of texture. It is a measure of the deviation of a surface in the direction of the normal vector from the surface as compared to an ideal. Measuring surface roughness is like measuring the waves in the ocean and how high and how frequently they occur.

Measurements are obtained from passing a profilometer slowly over the surface of a very small line of the finish surface. Usually several readings are taken, and an average is arrived at. Ra values help to narrow down the type of finish desired, but Ra has little significance when it comes to reflectivity of the surface.

TABLE 3.6 Descriptions of different finish categories used on stainless steel.

Operation	Description
Grinding	Removes metal. Takes out imperfections and changes thickness. Grinding always produces a high level of coarseness. The surface will exhibit resistance and friction when a hand or a microfiber towel is run over it. Very difficult to keep clean.
Polishing	Removes trace amounts of metal. Describes a finish such as No. 4. These surfaces can be wiped down with a fine cloth without its snagging on the surface. A No. 3 finish is a coarser version of a No. 4 finish.
Wet polishing	Removes trace amounts of metal. Finer and brighter finish obtained. This is a more refined version of a No. 4 finish; the surface is smooth to the touch and/or to the pass of a fine cloth. Easier to maintain.
Brushing	No metal is removed. Uses abrasives to texture surface. Fine lines. This is the Hairline finish applied by light brushing. Easier to clean and somewhat easier to repair than a No. 4 finish.
Buffing	No metal removed. Smoothing process involving pastes and rouge. This results in a mirror-like surface with few visible lines.

Other methods are used to measure surface roughness as well. Optical profilers obtain a significant amount of surface information, but a correlation between the light-reflecting quality of a stainless steel surface intended for art or architecture and the readings obtained from an optical profiler has yet to be revealed on a consistent basis (Table 3.7).

No. 3, No. 4, and Hairline Finishes

Directional satin finishes are common stainless steel finishes that have been used for decades to produce everything from countertops to column covers. It is interesting to note, though, that it is

TABLE 3.7 Surface roughness and grit size.

Grit size	Approximate roughness: Ra (μm)	Approximate roughness: Ra (μin)
500	0.10–0.25	4–10
320	0.15–0.38	6–15
240	0.20–0.51	8–20
180	0.64	25 max
120	1.14	45 max
60	3.56	140 max

not possible to match one of these finishes after years of being in place with newly finished sheets. These finishes have their own signature once produced; they are microscopically unique. The color and tone of the base metal, and the temper and hardness are just some of the parameters that affect the depth, quality, and reflective aspects of a directional satin finish. Matching these after a period of time is not possible. Up close they may be very similar, but the quality of the light reflection at greater distances and at angles will cause the surface appearance to be visibly different.

Each of these finishes is applied to coils, sheets, or fabricated shapes of stainless steel using various sanding belts. The belts can be aluminum oxide, silicon carbide, or Scotch-Brite™, a material made of aluminum oxide and titanium oxide impregnated nylon. A particular grit level is assigned to the belt, and this grit level will achieve the finish.

No. 3 finish	Coarse	60–120 grit
No. 4 finish	Medium to fine	150–320 grit
Hairline	Fine	600–800 grit Scotch-Brite

When an order is placed for any of the directional satins, the finish is applied to a coil of metal or to the individual plates of metal by mechanical means. The process involves passing a moving sanding belt impregnated with the abrasive over the surface of the metal while applying slight pressure. The metal is also usually moving. Both abrasive belt and metal move at a controlled rate. The abrasive creates linear scratch lines on the metal surface as it passes over it.

The wear on the belts is subject to the quality control processes of the finishing facility. Most change out the belts at a preestablished wear estimate. If this is not performed, the finish will be different as the belt wears. The type of belt, the type of grit, and the number of passes will determine the level of surface scratching applied. Grit stands for the particle size that is impregnated in the belt used to induce the finish. A roughness analysis is performed on the finished surface using a special stylus called a profilometer. A profilometer will read the average depth (Ra), maximum depth (Rmax), and uniformity (Rz) of the scratch produced. These will give a rough idea of how aggressive the abrasive is on the surface. They will provide a measure to determine the effectiveness of an abrasive pass but will not necessarily provide a qualitative analysis of the finish in different lighting conditions. Roughness is a good indicator of the level of maintenance needed to clean or keep a surface clean. The coarser the surface profile is, the more maintenance will be required to keep it clean.

The No. 3 Finish

The No. 3 finish is the coarsest of the directional finishes. On close inspection, the scratches are seen as short parallel strokes of variable depth. This finish can be blended around welds and corners using hand tools, but matching the finish in the center of a surface is not easily achieved. Matching one manufacturer or finishing house's No. 3 finish to another is not possible either. The No. 3 finish has deeper ridges induced from the coarse grit and there is a skipping appearance, with short, brighter ridges dispersed in the field. When passing your hand or a microfiber cloth across the grain produced by the abrasive you will feel the roughness. This finish is rarely used in architecture or art because of the roughness. The Ra reading of a No. 3 finish is no less than a 40.

There are numerous levels of No. 3 finish, two of which are shown in the following table.

Coarse No. 3 finish	100 grit	Ra target of 61	RMS target of 69
Medium No. 3 finish	120 grit	Ra target of 52	RMS target of 58

The No. 3 finish is used where cleaning is frequent, such as for interior surfaces or kitchen utensils. It is possible to blend this finish around welds and perform slight repairs to damaged surfaces. Cross-grain damage is the most common. This is where the tops of the ridges induced in the sanding process are taken down. Repair requires blending of the finish and matching the deeper grain and small skipping nature of the finish. This finish can be produced on a coil of stainless steel, and on stainless steel sheets, plates, tubes, and bars.

The No. 4 Finish

This finish is the most common directional finish used in art and architecture. It is similar to No. 3 in appearance, with finer parallel lines. Like the No. 3 finish, it is difficult to repair damages in the field of a flat surface. The No. 4 finish can be blended around corners and welds. No. 4 produces a good lustrous surface with a diffuse reflection in one direction and a higher gloss reflection in the perpendicular direction. Matching surfaces from different producers of the finish can be difficult to impossible. The finish is produced using aluminum oxide, silicon carbide, or Scotch-Brite pads. Each abrasive will give a slightly different appearance in certain lights. Each abrasive produces a different shape to the scratch profile (Figure 3.12).

The No. 4 finish has longer, parallel lines as compared to a No. 3 finish. The lines are finer and not as deep and variable as those in the No. 3 finish. The surface is easier to clean due to this lack of roughness.

The No. 4 finish can be both dry sanded and polished and wet sanded and polished. With wet sanding, the polish is produced under oils. This cuts the surface to produce a finer, more reflective grain. This finish can be produced on a coil of stainless steel and on stainless steel sheets, plates, tubes, and bars.

Coarse No. 4 finish	150 grit	Target Ra of 42	Target RMS of 34
Medium No. 4 finish	180 grit	Target Ra of 30	Target RMS of 22
Fine No. 4 finish	220 grit	Target Ra of 20	Target RMS of 18

Brushed Finish (Hairline)

The brushed finishes go by various trade names. A common name is Hairline.

Other names are also used to describe this finish, such as HL, Fineline, and Longrain, as well as several others, but they all essentially have similar qualities. The Hairline finish is distinguished by long, fine parallel scratch lines. The finish is much finer than the No. 3 and No. 4 finishes. It is

FIGURE 3.12 Weatherhead School of Management, Case Western Reserve University, designed by Frank Gehry. The finish is a No. 4 directional satin.

produced by passing a Scotch-Brite pad or similar abrasive pad or belt over the surface. Typically the finish is very fine and the grit used falls within a broad range: from 280 grit to as fine as 600 grit. Because of this, the finish is significantly more reflective and can have hot spots or glaring reflectivity similar to a mirror surface. A Hairline finish can be restored in the event of a mar or scratch with more ease than No. 3 and No. 4 directional finishes. Matching and blending of this finish is subject to the pressure of application of the abrasive pad, the coarseness of the abrasive pad itself, and the skill of the operator (Table 3.8).

The Hairline finish is frequently used as an interior accent finish or applied over a No. 2 or 2BA finish to produce a soft, lustrous finish. Due to the fineness of the finish, it is simpler to clean and maintain and does not hold fingerprint oils to the degree a No. 3 or No. 4 finish will. This finish

TABLE 3.8 Scotch-Brite pad and corresponding closest grit used in finishing.

Color of 3M Scotch-Brite pad	Comparable grit
Tan	120–150
Dark gray	180–220
Brown	280–320
Maroon	320–400
Green	600
Light gray	600–800
White	1200–1500

can be produced on a coil of stainless steel and on stainless steel sheets, plates, tubes, and bars. This finish is also applied to fabricated surfaces due to the beauty and ease of blending it around shapes.

NON-DIRECTIONAL FINISHES

Non-directional finishes are characterized by having a diffuse reflectivity. They are more expensive than directional finishes because they are typically produced a sheet at a time or a part at a time. These finishes capture the luster of stainless steel well. They can be applied to all forms of stainless steel sheet, plate, bar, tube, and casting. They reflect light hitting the surface in all directions. There is still a level of polarity in the light that is reflected from these surfaces, which can sometimes accentuate small surface irregularities. They can be repaired by blending the new surface into the older, damaged surface. Some of the more common non-directional finishes are shown in Table 3.9.

Angel Hair Finish

The non-directional finish Angel Hair® was developed while working with the great architect Frank Gehry. He wanted a finish that was more refined than a mill surface, a finish that would disperse

TABLE 3.9 Non-directional finishes.

Finish	Description
Angel hair/vibration	Small overlapping arcs
Glass bead	Small overlapping dishes
Stainless steel shot	Large, deep craters
Swirl/grind	Deep overlapping scratches

light while still possessing the luster and tone of a natural surface. Initially there were several dozen variations of the Angel Hair finish, each with subtle differences in reflectivity. The difficulty was to remove the mottling effect that could be seen when the surface was observed from a short distance. Overlapping patterns, such as scallops, would appear, or stripping patterns from the lapping process. Early Angel Hair surfaces have that overlapping scalloping, which appears in certain lighting. Some early versions and copies today are produced by hand. The result is a mottled inconsistent surface. There is no way around this when the finish is applied manually. Pressure differences and symmetrical movement of the discs used to apply the finish will create these ghosting effects. Up close all looks good; however, observe the surface from a distance and the scallop or lines appear (Figure 3.13).

True Angel Hair is a dual surface: one level of texture below another. The small circular scratches are induced in such a way that symmetry is disrupted, eliminating the stripping and scalloping effects. There are many techniques used today to produce the Angel Hair surface. Advancements in robotics have taken the finishing process in a new direction that maintains control and quality and eliminates the disturbing patterns created by low-quality production.

The main benefit of these finishes is the behavior exhibited by the interaction with light. The small arching scratches effectively scatter light and produce a diffuse reflection possessing different qualities when viewed up close or at a distance (Figure 3.14).

FIGURE 3.13 NASCAR headquarters, designed by Pei Cobb Freed & Partners.

FIGURE 3.14 Angel Hair finish in sun, and with an overcast sky.

Another major benefit is that they are repairable and easy to blend around welds.

In reality, that is where the finish was born. Repairing automobile dents, finishing welded sections of steel, and even the sanding of wood floors all use an electric sander that vibrates in tight swirls. This works well when finishing localized edges or preparing for paint and varnishes, but on stainless steel surfaces of size, or for a series of panel elements assembled to create a large area, this hand process produces a mottled, irregular appearance. Semiautomated machines were developed to produce the finish for architectural surfaces. The Angel Hair process uses a patented device that breaks the symmetry of the repeating arc.

Other systems use a sanding belt that moves as a sheet is passed below the belt. The result is more like small, snaking, curved lines than a series of overlapping arcs. The effect is similar when it comes to diffuse reflective appearance. This finish is sometimes referred to as Vibration.

The reflection generated by a diffuse, non-directional finish is markedly different from the reflection from a directional satin finish. True Angel Hair finish is a velvety, consistent, elegant surface. From sheet to sheet and edge to edge, this finish is like the threads in a fine suit (Figure 3.15).

FIGURE 3.15 The Petersen Automotive Museum, designed by Kohn Pedersen Fox Architects. The Angel Hair finish is 2 mm thick.
Source: Photo by Tex Jernigan | ARKO.

Glass Bead Finish

Stainless steel is well suited for glass bead blasting. The glass bead blast is a non-directional finish produced by impacting the surface of stainless steel with minute glass beads. These beads, striking

FIGURE 3.16 Glass bead blast.

the surface with significant energy, will impart minute, overlapping concave dishes across the entire surface of the stainless steel (Figure 3.16).

Glass bead blasting is not sand blasting. Glass bead blasting applies an inert material and does not contaminate the stainless steel or create undercuts that develop into areas where corrosion will be initiated. Minute, overlapping dishes or craters are induced by the impact of the glass beads. Glass beads are made from lead-free soda-lime glass. There is localized work hardening of the surface from the small impact craters produced by the high-energy process of blasting. The impacting causes the surface to go under tension. If the blasting is applied to one side of the surface only, it will shape thin stainless steel and cause localized convexity. To counter this shaping and balance out the stress within the sheets, the reverse side must be blasted to keep the sheet material flat. When properly applied, the resulting sheet is flat and residual stress is reduced.

Glass beads are available in many sizes. They are provided as small spheres determined by mesh size. There are several common sizes in use.

#13 (fine)	170–325 mesh	45–90 μm
#10 (medium-fine)	100–170 mesh	90–150 μm
#8 (medium)	70–100 mesh	150–212 μm
#6 (medium-coarse)	50–70 mesh	212–325 μm
#5 (coarse)	40–50 mesh	325–425 μm
#4 and #5 (extra coarse)	20–40 mesh	425–850 μm

For glass bead blast finishes produced for the art and architectural marketplace, the coarse finishes are rarely used. Fine and medium glass beads at a specific pressure are used on stainless steel.

Systems designed for glass bead blasting a surface apply the glass beads evenly and remove most of the shards created in the blasting process. This is critical for producing a good glass bead blast surface. The shards are removed in a venturi, and the pressure of the beads hitting the metal is accurately controlled. The glass beads are sent through pressurized jets of clean, dry air. As the glass beads impact the surface of the stainless steel, they transfer their kinetic energy to the surface, slightly stretching the metal like thousands of small ballpeen hammers. Metal is not removed in the process. Some of the glass fractures into shards. These must be removed or they create sharp indentations on the surface that can become undercuts or coarse regions where dirt can catch hold. The shards are removed as the glass beads are suctioned up out of a hopper that collects the spent glass beads.

Glass bead finish can be applied over other textures to mute reflectivity, or the finish can receive chemical treatments to add color to the matte finish. Decorative surfaces can be created with glass beads by masking. Glass bead finishes are diffuse satin finishes. They are low in relative gloss (but not as low as an Angel Hair finish). The finish can be described as crystalline when in bright sun and almost pewter-like in when the sky is overcast. Glass bead surfaces that have been electropolished resemble new tin coatings: bright, but diffuse in reflectivity. Glass bead finishing can be applied to sheet, plate, bar, and tubing – virtually any form of stainless steel.

Glass beads will not clean the surface of a stainless steel sheet. If fingerprints are on the sheet before blasting, they will still exist after blasting and be more difficult to remove. New glass bead surfaces are very porous and will readily fingerprint. Allow the sheets to sit exposed to the air for a short time period before applying a protective plastic film coating. Be sure there are no air bubbles in the plastic film coating. Once the sheet has been formed and installed, remove the coating and allow the air to reach the surface. Moisture in the air will collect on the surface and fingerprinting will be less apparent. The surface finish can be cleaned with glass cleaner and other mild solvents. Glass bead surfaces receive anti-fingerprint treatments well, but there is a slight tone change that occurs when the thin anti-fingerprint coatings are applied.

Improper blasting or hand blasting will produce a mottled, uneven surface. Thin sheets can be warped and bowed from the energy imparted from blasting if the surface is not equally treated on the reverse side. Special CNC (computer numerical control) or reciprocating machines are capable of producing the finish evenly and consistently. The finish is induced on both sides and the glass beads are collected as the shards are removed. Oil or water transferred through the air blast will stain the finish and cannot be cleaned or repaired, so it is important to have the necessary condensers and filters in good working operation.

Other blast media exist but should be questioned when applied to stainless steel. Never use sand. Sand is too coarse and it is nearly impossible to ensure that it is clean and free of substances that can lead to degradation of the surface. Garnet also cuts the surface and makes a coarse texture that is subject to holding contaminates on the surface. Steel shot will contaminate the surface of stainless and require extensive chemical treatment to facilitate removal. Steel shot and steel cut wire should never be used on stainless steel. Besides glass, the only other material suitable for stainless steel blasting in order to provide a finish is stainless steel shot.

There have been studies on the corrosion behavior of glass bead blasted stainless steel. The studies examine the roughness coefficients of various finishes and compare electrochemical parameters of S30400 stainless steel of various finishes in sodium chloride exposure tests. The result is a comparison of the potentials for crevice corrosion on the various finishes tested: glass bead, glass bead plus electropolish, sand blast, directional ground finish (No. 4), and structure rolled (coined).

The problem with many of these studies is the parameters used to evaluate the glass bead application. In practice, glass bead finishes perform very well on many exposures. Proper glass bead

FIGURE 3.17 IBM headquarters, designed by Kohn Pedersen Fox Architects. The glass bead stainless steel finish is 2 mm thick. The close-up image shows the remarkable surface after more than 20 years of exposure.

blasting must remove the shards of broken glass to prevent undercutting. Once the finish is applied, exposed to the atmosphere, and allowed to develop the chromium oxide layer, the surface will perform just as well as the directional finishes, without electropolishing. There are decades-old installations of glass bead stainless steel that look as if they were newly installed.

The glass bead surface has depth in reflectivity. Images are misleading; they show a flat surface, whereas in reality a glass bead surface reflects the surrounding light and glows like a crystal in the sunlight and looks like pewter in shadow (Figure 3.17).

The Stainless Steel Shot Blast Finish

Finishing a surface with stainless steel shot produces a work-hardened surface that is very resistant to abrasion and marring. The surface is virtually scratch resistant for most common materials due to the increased hardness induced by selectively cold working the surface. Stainless steel shot blasting produces a finish similar to glass bead blasting, only far deeper and more pronounced. The concave craters are larger and more defined. The surface achieved is brighter and harder than the surface produced by glass bead blasting. One challenge, though, is shaping. A surface blasted with stainless steel shot will shape extensively. Reverse side blasting is necessary if you wish to arrive at a flat sheet of metal. The hardened sheet or plate is more difficult to work. The energy imparted is significant and the equipment needed to produce the finish will require extensive upkeep. The act of blasting with stainless steel takes a toll on the equipment utilized and the wear on nozzles can change and alter the surface's appearance.

It is not a finish commonly used in art and architecture. It is somewhat limited to custom-shaped tubes or bars; on sheets and plates, not only is the surface harder but it does make it difficult to work the material. For areas of very high impact, there are few finishes that will perform as well.

CUSTOM MECHANICAL FINISHES

There are innumerable custom finishes produced by artists and custom architectural fabrication facilities. Since many of these are produced by manual or semimanual means, they are limited in terms of commercial viability. However, it is important to discuss the potential and the commonalities of these finishes (Figure 3.18).

A simple grinder in the hand of a craftsperson or artist can achieve beautiful ground finishes that reflect light in such a way they can produce a lively, visual assault that changes from light to dark as viewpoint changes. Adjusting the way light reflects off the surface can create interesting polarized reflections. Masking to create artistic changes and combining mechanical and chemical finishes can create interesting effects on the viewer (Figure 3.19).

These coarse finishes are produced by grinding the surface at different angles. They are applied usually with a handheld grinder brushed over the surface. Each subsequent brush overlaps another ground surface in a lapping sequence. The result is a surface that interplays with light reflecting off

FIGURE 3.18 Examples of custom mechanical finishes.

the individual ground marks. Artistic uses of this finish give a three-dimensional quality as light and dark reflections produce a contrast, an illusion of high and low regions that change as the surface is viewed from different angles. This finish can be put on sheet, plate, rod, and tube. The effect can bring a surface to life as the light reflects from the ever-changing angles produced by the grind marks.

FIGURE 3.19 Custom finishes applied to art. (*Left*) *Sprout*, by Beth Nybeck; (*right*) *Cubi XXVI*, by David Smith.

HIGHLY CUSTOM SURFACE FINISHES

There are two custom finishes that merit discussion because they are not only unique but capable of being repeated in scale. These are blast striping and crosshatch finishing.

Blast striping involves masking sheets of stainless steel and passing them through a glass bead blasting machine to produce one finish over another. This can be done on interference colored stainless steel as well. An example of this is the 9/11 Memorial & Museum in New York. Here the designer, Snøhetta, worked with the finishing company to achieve a subtle blending of stripes across the surface of the building. Different base metal surfaces were produced: No. 8 mirror, No. 4, and a softer No. 4. The surfaces were carefully and accurately masked, then the No. 8 and No. 4 finishes were blasted with glass (Figure 3.20).

FIGURE 3.20 National September 11 Memorial Museum, designed by Snøhetta Architects.

The crosshatch finish, whose commercial name is CrossFire™,[3] is a No. 4 finish applied at an angle, then rotated and applied at the opposing angle. After the finish is applied it is electropolished and colored blue by the light interference method. The result is a striking surface for the Contemporary Jewish Museum in San Francisco. Different angles can be selected and induced into the surface. The designer, Daniel Libeskind, chose a 60° angle to work with the pattern of the panels. The appearance changes as the angle of sunlight hitting the surface changes and reflects back to the viewer.

The interference color adds a further dimension of beauty to the surface (Figure 3.21).

[3]CrossFire is a custom finish produced by Rimex Rigidized Metals Ltd., Enfield, England.

FIGURE 3.21 Contemporary Jewish Museum in San Francisco, designed by Daniel Libeskind.

MIRROR FINISHES

Mirror finishes are by nature non-directional. There are several levels of mirror finishes. All are specular, one-to-one reflective polished surfaces where the angle of incident is equal to angle of reflection. A mirror surfaces is produced by successive grinding and polishing of the surface. Mirror polishes have different levels of quality whose ranking depends on the visibility of polish lines. For example, a No. 7 finish is considered acceptable if fine polishing lines are apparent, whereas a Super No. 8 will have no apparent lines. Producing these finishes requires specialized equipment and skills. Removing the grit lines requires polishing skills and proper equipment. The process of achieving high-quality mirror surfaces requires using rouges and fine polishes in succession to achieve a high polish.

No. 7 and No. 8 Mirror Finish

These are highly reflective finishes produced by progressively finer abrasive polishing and then finished with buffing. A No. 7 finish has visible grit lines when viewed in such a way that light strikes the remaining cutting lines. This finish is easier to match and repair because of the fewer steps involved with producing it. Many hand-polished and buffed surfaces fall under the No. 7 classification because of the appearance of these fine lines that are visible after the buffing. A No. 8 finish can be achieved with more buffing.

A No. 8 finish is polished to a minimum of 240 grit before buffing operations. Buffing involves a step known as cutting, in which the grit lines are taken down using various buffing compounds applied to cloth wheels moving at a high speed. The second step in buffing is called color buffing. This takes the surface further by applying rouges that are progressively finer (Figure 3.22).

Both of these finishes have visible grit lines, with those on the No. 7 being more pronounced in various lighting conditions. The No. 8 grit lines are visible in fewer lighting conditions and from fewer lighting angles. These finishes are both specular finishes. They can be repaired if scratched or marred by following steps similar to those taken to produce the finish initially.

Super Mirror Finishes

Super Mirror finishes were developed commercially in Japan in the mid-to-late 1980s. Since then they have been moving to replace the No. 8 finish of old. Essentially these finishes are free of grit

FIGURE 3.22 A No. 7 finish on the Dream Downtown Hotel in New York City, designed by Handel Architects.

FIGURE 3.23 Examples of No. 8 and Super Mirror finishes.

lines or even remnants of polishing lines. They very closely resemble the reflective quality of a true glass mirror. These finishes are produced similarly to a No. 8 finish, except further steps are taken to remove all indications of haze or the small scratch lines visible in a No. 8 finish. This finish can be chemically treated to produce color by the interference method or by PVD. Adding color will result in a surface with a highly reflective, deep, metallic color.

This high-quality mirror finish is produced on very flat sheets with a limited thickness range. Thin sheets that are less than 1 mm thick are difficult to polish to a Super Mirror finish. Thick sheets that are greater than 3 mm do not polish well either. The grain quality is not as good as the thinner cold rolled sheets (Figure 3.23).

Placing a Super Mirror reflective surface on tube or bar can be accomplished using rouges and high-speed buffing equipment. The process is slow and expensive. It requires skill and patience.

PLASTIC DEFORMATION SURFACE FINISHES

These finishes are all characterized by inducing enough force into the stainless steel sheet to impart permanent plastic deformation. Typically, these are patterns cut into cylindrical dies. As the metal is passed between the cylinders, sufficient force pushes the stainless steel into the pattern cut into the roll. The metal yields under the force and the pattern is permanently set into the stainless steel.

Other methods use stamping dies, in which the stainless sheet is pushed into a mold. The use of pliable bladders to protect the surface of the stainless steel is often incorporated into the process. In this method the durometer[4] (the measure of the hardness of plastics and rubbers) must be set correctly so that it has the ability to impart the force into the stainless steel. Some patterns use only a female die and the stainless steel is pushed into the opening. The metal yields and deforms into the opening in a dish-like formation.

These are highly custom surfaces created by pushing thin stainless steel under extreme pressure. The difference between this surface finish and that of the embossed or selective embossed finishes involve overcoming the internal stress in the metal and pushing the material into deep, out-of-plane dies. Plastic deformation in the metal occurs as the stainless steel conforms to the die. These finishes are characterized by macro-scale surface undulations that provide the overall surfacing appearance. Essentially these surfaces are produced by pressing the metal into large die blocks. Critical challenges center around bends or returns where the deformation and work hardening create conditions where the stress is not consistent.

"WATER" FINISH

This is an example of a deep-pressed stainless steel surface. The die is a shaped surface milled into a block of rigid material. The stainless steel is pushed into the surface under extreme pressure using a specially designed durometer rubber. The surface creates a ripple appearance that reflects light in a way comparable to the surface of water. The depth of the mold and the pressure to shape the stainless steel must be sufficient to overcome the springback and yield point of the metal. This finish utilizes the highly reflective luster of stainless steel. As you move around and view this finish it shimmers, as light reflects off the undulating surface back to the eye like a light ripple on a calm sea. Contrasting light and dark regions are apparent as well (Figure 3.24).

Another means of producing a similar finish is by passing the metal through a roll with patterns induced and using rubber mats to press the stainless steel into the surface. This can expand the potential for producing highly custom finishes with a tactile character (Figure 3.25).

EMBOSSING

Embossing is a process in which stainless steel is rolled through matching sets of pattern rolls, where one roll has the pattern cut into it and the other roll has the pattern protruding outward. The patterns on the rolls are set in alignment, and as a thin sheet of metal is passed through the rolls, the pattern

---

[4]Durometer is the term given to a scale that measures a material's resistance to indentation; it is usually applied to rubbers and plastics.

FIGURE 3.24 Custom "Water" finish on a mirror-finish stainless steel, perforated stainless steel, and interference blue stainless steel.

is imprinted on the surface with one side of the sheet matching the inward pattern and the other side the outward pattern on the other roll. There are many patterns produced on thin stainless steel sheet in this way. The process work hardens the metal and thus equalizes the differential stress that might exist in a thin sheet. The result is a flat surface of metal with a pattern imprinted from edge to edge.

This process is best suited for thin sheet stainless and can be applied to coils. The maximum thickness that can be worked is 2 mm (0.078 in.); beyond that little patterning is produced.

Embossing serves several purposes. Embossing allows for thinner sheet material to be used for applications where waviness will impact the end aesthetic appearance. Embossing can achieve better utilization of mill coils by reducing quality concerns. Embossing removes the edge wave and surface wave, and conceals defects that might otherwise lead to rejection.

FIGURE 3.25 Column covers made from custom roll-embossed stainless steel.
Source: Photo courtesy of Rimex Metals.

Embossing can be further finished by subsequent polishing, creating highlighted tone patterns on the sheet. Concealing scratches and mars, these finishes are excellent choices for high-traffic areas where abuse or mishandling is common. Strength and corrosion resistance are not altered by embossing.

Embossed textures can be applied to sheets that have any of the mechanical or chemical finishes. They can also be post-finished or "highlighted" to develop surfaces unique to embossed sheets. Embossing can be applied to continuous coils and individual sheets (Figure 3.26).

A limitation of embossing is that the patterns must go from edge to edge on a sheet. Patterns induced onto the surface are relatively small grids made up of indentations that are usually less than 10 mm across. The depth of the pattern is shallow and no more than 1.5 mm out of plane.

SELECTIVE EMBOSSING

Selective embossing uses computer controlled equipment to selectively upset the surface of stainless steel sheets or plates. They require concentrated pressure and specialized dies to achieve selective embossing. Embossing this way is significantly slower, but the ability to create unique, even one of

FIGURE 3.26 Embossed stainless steel on a Neiman Marcus store, designed by Elkus Manfredi.

a kind, surface finishes sometimes coupled with perforations can offer a designer an endless canvas to work with.

Selective embossing, or "bumping" as it is sometimes referred to, can go inward or outward from the face. Embossing this way can yield any number of geometric shapes or custom shapes. Localized work hardening around where the embossing has been placed can warp the flat sheet or distort the surface. Tests should be performed to identify the correct temper and thickness to meet the design criteria.

FIGURE 3.27 Embossed and perforated stainless steel on the Fairmont Pacific Rim hotel, designed by James K. M. Cheng Architects.

Selective embossing can be performed on any of the sheet finishes: mechanical, chemical, or both. Like embossing, post-finishes can be performed on the tops of embossed surfaces to change the reflectivity and expand the design potential (Figure 3.27).

Limitations are in the die size. The dies are normally 75 mm or less, so a selective embossing element must fit within this circle. There are larger dies that are as much as 150 mm in diameter, but they are rarely used. The depth of a selective embossed element is shallow. Out-of-plane depths are less than 3 mm.

Another limitation is clearance from emboss to emboss. Since the process is a localized stamping process, placing the embossed element too close to another element can smash a portion of a previous placed bump. Spacing for clearance is an important consideration when performing selective embossing.

COINING

Coining is a process where the sheet or plate material is passed through very high-pressure rolls and one roll has a pattern cut into the roll while the other roll is smooth. As the metal is passed through

FIGURE 3.28 Linen finish on a building in the Hudson Yards development, designed by Kohn Pedersen Fox Architects.

these rolls the pressure pushes the surface into the cut pattern. The result is that one side of the metal has an outward protruding pattern that is the negative of the cut pattern on the roll.

This process is only possible on sheet or plate material that is a minimum of 2 mm (0.078 in.) thick. The pressure is significant, so wear on the rolls needs to be addressed frequently. The metal surface is work hardened, so it improves impact and scratch resistance.

Subsequent polishing of the surface will highlight the raised portions of the pattern, allowing for further appearance modifications to the surface. This process is typically performed on mill surface finishes; in particular, the 2B and 2BA mill finishes.

The process can hide many defects that are common in mill surface material, such as streaking and slight chatter (Figure 3.28).

The limitations of coining are similar to those of embossing. The pattern must run from edge to edge on the sheet. The pattern is small and shallow. Coining patterns are less than 1 mm in height.

CHEMICAL FINISHES: COLORING OF STAINLESS STEEL

Another fascinating attribute of stainless steels is their capacity to have deep metallic colors induced in them. The colors are not pigments applied to the surface in the way of paints or lacquers.

However, the application of semitransparent lacquers is possible and when correctly applied can use the high reflective character of the metal to create beautiful tones of color. With this technique color is achieved by pigments, and the coloring of stainless steels using resins and varnishes is not much different from the method used for aluminum or steels. It is the application of a coating that provides the basis for the color. Here we will discuss techniques uniquely connected to stainless steel: colors created from thin oxides and physical interaction with the stainless surface. There are two major commercial methods of adding color to stainless steel. Both give excellent results when correctly understood and executed. The first is known as interference coloring and the other is known as PVD coating.

Stainless steel can develop a thin oxide that will impart interference colors by one of three methods:

- Thermal oxidation
- Chemical oxidation
- Electrochemical oxidation

Alternatively, the PVD process can be used to impart color into stainless steel. In PVD coatings very thin layers of other metals or compounds are deposited atom by atom onto the surface of stainless steel. One form of PVD commonly used on architectural stainless steel surfaces is known as sputter deposition.

Thermal Oxidation Coloring

Thermal oxidation coloring involves heating stainless steel to develop a heat tint on the surface. It is augmented at times with chemical treatments that react with the chromium on the metal's surface. It is a slow process involving manual control and adjustment of the induced colors through intense heating and rapid cooling of the metal. The colors created are variable and uneven. As it thickens, the oxide develops into interference bands radiating out from the heat.

Once completed, the surface should be sealed with wax or a clear lacquer. It is somewhat fragile, and the metal surface can corrode if exposed to mildly corrosive environments.

Interference Colored Stainless Steel: The Inco Process

The Inco process, developed and originally patented by Inco Limited in the early 1970s, uses chemical acid baths to develop a thickened chromium oxide layer. An electrochemical process is used to harden the film and make it abrasion resistant. This thickened layer produces color by the phenomena of thin-film light interference: thus the name "interference coloring." The phenomena of light interference that induces the colors in the interference coloring process is directly attributed to the clear chromium oxide layer and its effect on the light wave passing though the film and back to the viewer. There are no pigments, inks, or dyes used to create the color (Figure 3.29).

FIGURE 3.29 Black Granex™ stainless steel walls.
Source: Photo courtesy of Rimex Metals.

The stainless steel alloy S30400 is commonly used for interference coloring and seems to develop the colors with more consistency and clarity. The alloy S31600 can be colored as well, but the clarity is not as good. S4xxxx stainless steels with high chromium content can develop deep colors as well.

Coloring of stainless steel is typically a sheet process and is limited by the size and width of the coloring tanks or the vacuum chambers used in physical vapor techniques. If they can fit in the coloring tank or chamber, other forms can also receive color. Three-dimensional forms may have variations from one surface to the other, but they can be colored. The Inco process is a chemical immersive process. Sheets are immersed in baths of chemicals to produce the color. If air is inside the part being colored, it will float and make producing the color difficult if not impossible, as when anodizing aluminum. The parts must drain and not hold acid in seams or joints.

Color from Light Interference

The phenomena of light interference involve two beams of light of the same wavelength travelling at parallel paths. One wave can interact with the second wave to produce what is known as either constructive reinforcement or destructive cancellation (Figure 3.30).

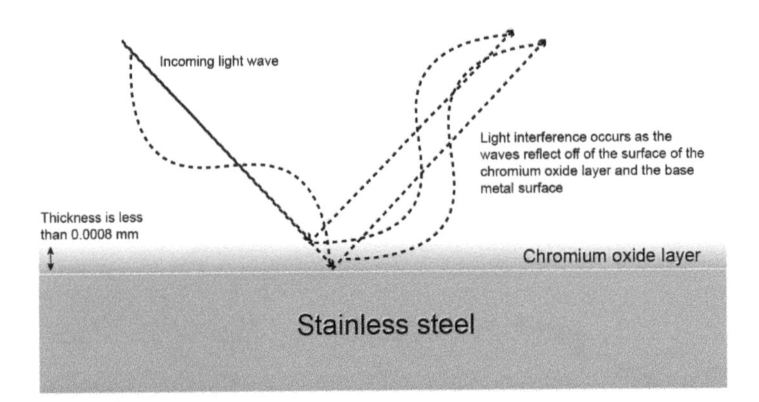

FIGURE 3.30 Light interference phenomena.

In the case of stainless steels, this condition can be induced by thin-film interference. An example of thin-film interference occurs when a thin layer of oil on water displays iridescent colors on the surface. The colors come from light interference caused by the thin oil film as the light passes through the oil and reflects off the surface of the water and light reflects off of the oil surface. The light waves interact in such a way as to generate the appearance of color. There are no pigments or dyes involved. The color is induced from the interaction of the two waves of light (Figure 3.31).

Coloring stainless steels through light interference was known early in the development of the metal. Chemical processes and thermal processes could create colors in somewhat predictable ways, but these colors were fragile and could be easily removed with the lightest of abrasions. The Inco process set out methods of controlling the color formation in a more predictable manner, but it is also a method of hardening the surface electrochemically. The initial film that develops is soft and porous. The Inco process hardens this film by making the colored stainless steel sheet a cathode and precipitating chromium oxide into the pores. This increases the density of the film without affecting the color. In analysis, the oxide that develops has a spinal structure enriched in chromium.

One of the first large-scale projects to use stainless steel colored using the Inco process was the Reiyūkai Shakaden temple in Tokyo. The roof is clad in over 18 500 square meters of a blue-black stainless steel. The interior ceiling has over 4800 square meters of gold stainless steel. Japanese designers and the Japanese stainless steel industry have embraced the metal coloring process and have created everything from colored stainless steel bathtubs to building surfaces. Japanese artisans have also developed ways to etch away areas and create fascinating artwork.

In Europe and the United States, the Inco process was a little slower in development. The first major projects in the United States were the Museum of Science and Industry in Tampa, Florida by Antoine Predock and the Team Disney Anaheim building in Anaheim, California by Frank Gehry. In Paris, the first buildings of the new Euro Disney Resort (also designed by Frank Gehry) were clad in colored stainless steel. Today you can find colored stainless steel in many commercial buildings and even in residences (Figures 3.32 and 3.33).

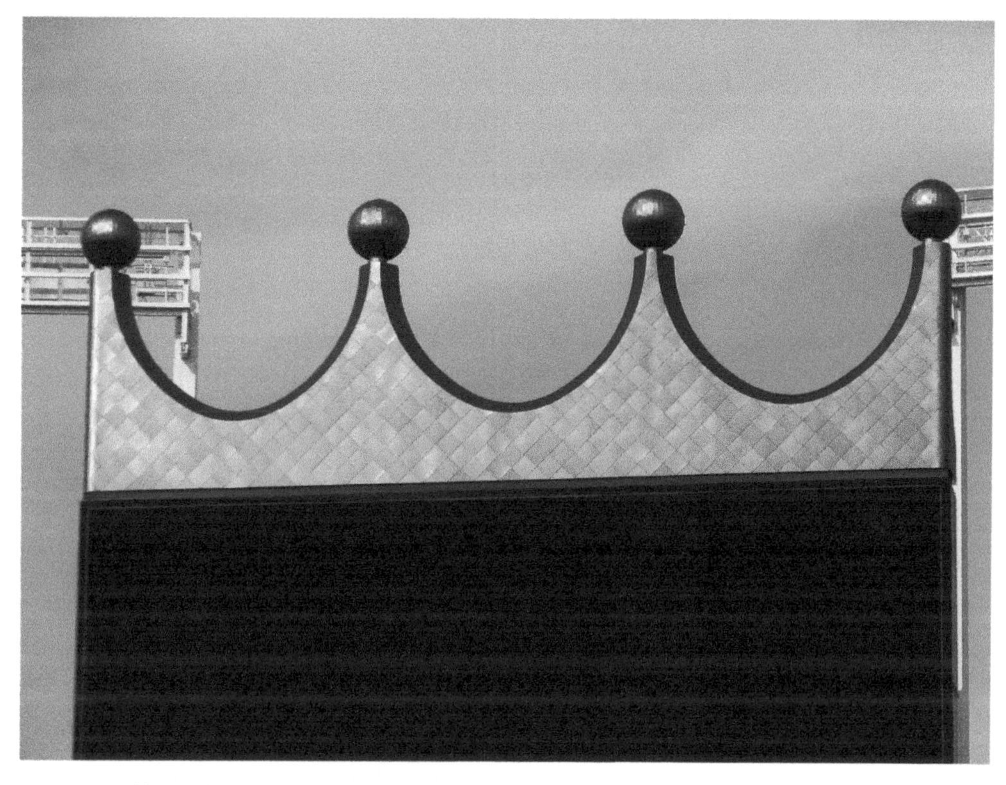

FIGURE 3.31 Gold stainless steel on the Kansas City Royals stadium scoreboard.

The Process of Creating Color

With stainless steel, we are dealing with an extremely thin layer of oxide. The natural coating that develops on stainless steel is 3–5 nm in thickness, or about 15 layers of atoms. This oxide normally is an iron-chromium oxyhydroxide.

The natural chromium oxide layer is enhanced by immersion of a clean stainless steel surface in a hot bath of sulfuric acid and chromic acid. The acid bath causes a chromium-rich oxide layer to grow. The thickness is mostly dependent on length of immersion. The clear chromium-rich oxide layer increases in thickness from 5 nm to as much as 360 nm.

Each step of the growth of the clear oxide corresponds to a selective interference of two reflected light waves. Due to the interference shift, portions of the wavelength are enhanced while other portions are canceled out. As the oxide grows, colors from interference develop. First a bronze color tone develops. This occurs with a thickness of approximately 20 nm. There is a very narrow range of bronze color tones, from a very light tone to a deeper bronze. Next, a range of blue tones develops: again, a very narrow range from a deep blue to a blue-black color. This is followed by a range of golden tones: first, a slight greenish tint, then a golden tone, then a rosy-gold tone. The next color is

FIGURE 3.32 Museum of Science and Industry in Tampa, Florida, designed by Antoine Predock.

FIGURE 3.33 Green stainless steel on the Team Disney Anaheim building, designed by Frank Gehry.

a short range of magenta, which is quickly followed by greenish color. The green has a thickness of approximately 360 nm.

In the process, the thickness of the chromium oxide is measured. Through tests, a thickness correlation to a particular target color is established. The thickness is determined by measuring the voltage resistance; this resistance corresponds to a certain thickness of the oxide film. As the thickness grows beyond the green color, however, the film begins to break down and the surface can deteriorate.

It is important to note that attempting to determine the color by visually looking at the metal as the process proceeds is not practical. The bath itself makes it difficult to see the color and it is only in subsequent rinse cycles that the color starts to emerge. Measuring the electrical potential and adjusting for tonal differences due to alloy constituents is the method used (Figure 3.34).

The color that our eye sees is dependent on the thickness of the clear oxide film and the reflective base finish on the stainless steel, as well as the angle of view. It is very important to understand that the color is not a pigment or dye. Therefore, what the eye is seeing is one of what are known as Newton's interference colors. You cannot mix them and arrive at some combination. The color is subject to the natural laws of physics and the ability of the company producing the color to predict the thickness and corresponding color that develops. The oxide film develops slowly at first. At about 20 nm the color is a grayish bronze, which corresponds with Newton's findings. As the film thickens

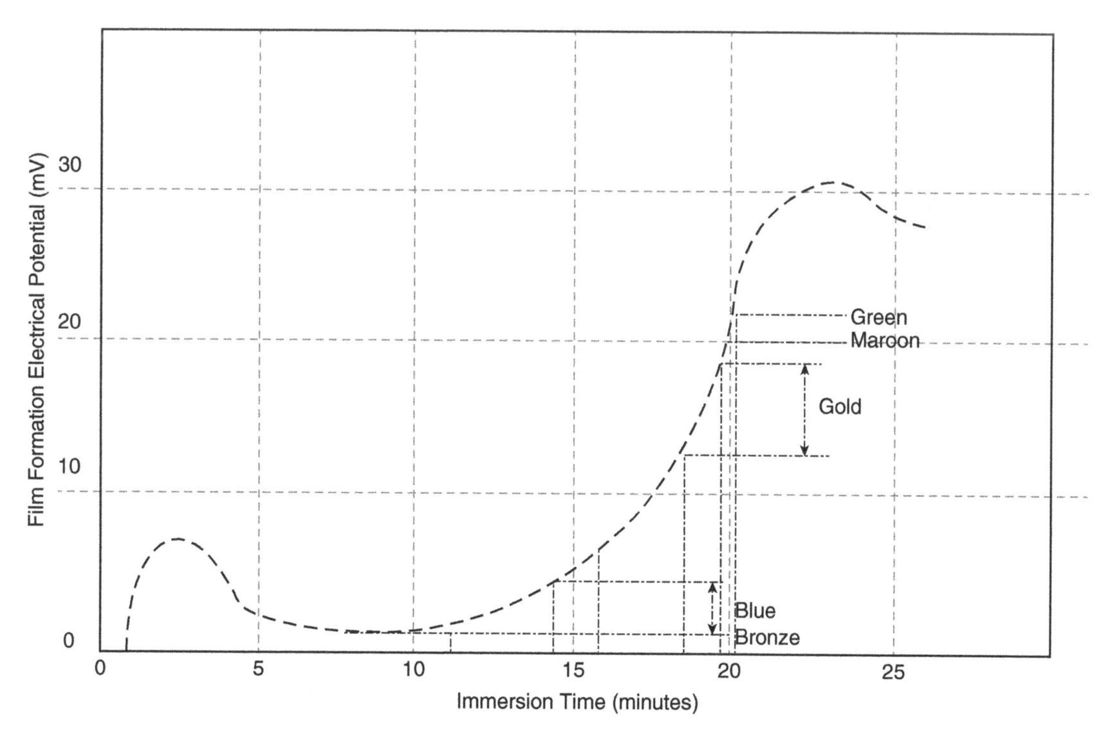

FIGURE 3.34 Coloring process: immersion time plotted against resistance.

beyond 50 nm, blue dominates, and as you move up the curve, the blue gets darker, almost black. Then the surface goes into a range of gold hues, followed quickly by a maroon color (some refer to this as red), then green. Green is the thickest oxide film. Beyond this thickness, the color goes drab and the film actually can weaken.

Gold stainless steel can have tints of red and even green depending on the angle of view and the base metal finish. This is due to the angle of view where the reflection takes light through a thicker clear oxide.

In the graph shown in Figure 3.34, maroon and green are very close because they are merely seconds apart in the color development process. There is no real range in color. They are more difficult to produce and they are more susceptible to change in chloride environments. There have been instances where red and green colors have deteriorated or become unstable from chloride exposures, both from deicing and from seaside environments. Other colors perform without issues in the same environments.

The bronze is in actuality a darkening of the surface that occurs. There is a bronze range: the color initially is a bronze tone, then a light bronze tone appears.

The initial buildup of the oxide is a reaction to the acids. There is an anodic dissolution of a portion of the surface as iron goes into solution. This is followed by a cathodic reduction of chromium anions as they develop into an oxyhydroxide. The film builds back rapidly in a chromium-rich oxide, which creates the interference colors. Other factors that have an effect on the color and consistency obtained are the nature and concentration of the bath, the alloy of stainless steel, the temperature consistency of the bath, and the stainless steel surface finish.

After the corresponding thickness of the chromium oxide layer has been established, the surface needs to be stabilized. Inducing color into stainless steel was studied as far back as 1927, when Dr. Hatfield, a British metallurgist from Sheffield, England, experimented with coloring stainless steels. The difficulty was stability. The film that produces the color from light interference is very porous and prone to damage. A hardening treatment is performed whereby the colored sheet is immersed in a bath of dilute phosphoric acid and chromic acid. A current is passed through the bath such that the colored sheet is made the cathode. The pores are filled and sealed with a hydrated chromium oxide. The final oxide layer has a spinal structure that is hydrated and enriched with chromium. This stabilizes the surface and hardens the oxide layer. This last step is required for all colored stainless steel surfaces. This process of controlling and predicting the color was developed by Inco.[5]

Adjusting the Color

The interference coloring process is not tunable like pigments in paint colors. You cannot achieve an intensity (for example, light blue or dark blue, or deep red or pink). Colors are specific to the light wave and the stainless steel color itself. For bronzes, blues, and golds there is a very narrow range of possibilities and control is difficult. You can "adjust" the color only slightly, either by removing color in post-processes, texturing the surface to alter reflectivity, or by altering the base finish.

[5]There were numerous patents for the coloring of stainless steels. Inco is credited with patenting the process of controlling the development of specific colors on chromium-bearing steels using electrical potential.

FIGURE 3.35 Textured and colored stainless steel, designed by Helix Architects.

Once the part or sheet has received the color, you can selectively remove the color by selectively polishing the oxide layer off, by applying a resist and removing the color, or by simply lightly repolishing the sheet or part (Figure 3.35).

Adding texture will alter the way light hits the surface. Bright spots may appear next to recessed areas on a highly localized basis. This alters the perceived color. You can polish the tops of embossed surfaces, removing the color and exposing the base metal on the peaks. Another approach is to adjust the base metal finish. A satin finish, such as Angel Hair or glass bead, will give a soft, satin appearance. A No. 4 finish will give a directional bright finish and a No. 8 or 9 finish will produce a beautiful colored mirror surface.

Applying an additional coating over the surface destroys the light interference effect. The colors are striking and beautiful, but they have their basis in natural phenomena that are subject to the laws of physics.

FIGURE 3.36 Resist applied over stainless steel, designed by Jan Hendrix.

You can remove the color on a selective basis. Similarly to removing the color tint on a weld, the oxide layer can be selectively dissolved and returned back to the original color. You can apply resist by means of silkscreen or photoresist (Figure 3.36).

It is important to note that an exact match in color from one sheet to the next is difficult, if not impossible, for certain base finishes. Very refined base metal finishes, such as mirror finishes that have been polished and cleaned, tend to yield the best results. Satin finishes, which are finishes that scatter light, can have more variability in color tones from one element or sheet to the next. It is somewhat easier to get good consistency across a surface with blues and golds than with reds and greens. Returning to the structure and trying to match a color produced at a previous time cannot be accomplished with any certainty. There are many variables in the process and controlling them all to a very tight specification is very difficult, if possible at all. The base makeup of the alloy itself will have an effect on the color achieved. Subtle variations in alloy consistency at the surface will affect the makeup of the chromium layer.

Coloring welded parts can be a challenge. The weld alters the metallurgical makeup of the surface. Keeping the weld to a minimum, removing all excess weld material, and grinding and polishing the surface will reduce the appearance of the color differences at the weld. Using the same alloy wire or fusing the stainless steel using TIG (tungsten inert gas) welding followed by passivation, light grinding, and polishing will yield the best results.

Producing Black Interference Color

Black interference is produced in ways similar to the standard coloring process. The black coloring process involves creating a surface oxide of a specific thickness where light reflecting off the dual surfaces aligns the wavelengths such that there is destructive cancelation of the light wave. Black can be a striking, rich color. It uses a distinct acid bath mixture.

Using the Inco process described for the other colors, a deep blue-black can be obtained. Another method incorporates a bath of sodium dichromate solution. The result is a charcoal black in S30400 stainless steel or a deep black in S41000 stainless steel.

Thermal Absorption of Dark Surfaces

Stainless steel is a poor conductor of heat. It tends to hold it. Therefore it is great for cooking utensils but distortions from welding are a challenge. Adding color to stainless steel increases its capacity to absorb certain wavelengths of light. This converts to heat. Stainless steel colored dark blue is used in solar collectors because of the color's ability to absorb as much as 90% of incident light.

Using colored stainless steel in places where sunlight will be intensely absorbed and where it could be touched should be avoided due to the heat-absorption characteristics the various colors induce into the metal.

Physical Vapor Deposition (PVD) Coloring

PVD is another way of producing remarkable colors on stainless steel. PVD is a name given to a process that involve vaporizing a material from a solid form and transporting the vaporized material to a metal surface where it condenses and forms a thin coating on the surface. By using reactive gases such as nitrogen, oxygen, or acetylene in a low-pressure plasma chamber, compounds of carbides, nitrides, and oxides can be deposited onto the surface of stainless steel.

All PVD coatings are thin, even those that are be built up as the operation continues. For most artistic and architectural purposes, the coatings are thin decorative films. When applied on stainless steel, the luster and finish of the metal enhances the color. PVD came into use in architecture in the late 1980s. Its use on small hardware parts, such as faucets and door handles, is replacing the electroplating process. It has better hardness and is an environmentally clean technique, as there is no waste product needing disposal to speak of. PVD is a dry process and does not create effluent to be treated, nor does it involve gases or volatiles that have to be collected. It does require high electrical current and involves heat.

PVD has become a viable coloring technique in the art and architecture market, as development and process improvements have made it more cost effective. One constraint to the application of the film is line of sight as the coating is applied. In other words, off sets and undersides do not see the film development. There are ways to accomplish this in small parts and

FIGURE 3.37 Miami Design District parking garage, designed by Leong Leong Architects. The laser-cut façade of Angel Hair stainless steel is colored using PVD.

hardware, but sheets are typically coated on one side, with the other side remaining uncoated (Figure 3.37).

Welded surfaces can be coated with thin film. The coating covers the weld with the same thickness of film as the adjoining metal. The film is not a concealing film because it is only a few atoms thick (2–6 μm in thickness), so mirror, Hairline, Angel Hair, and Vibration stainless steel finishes all can be coated with the thin films of vapor deposition. Sheets coated with the PVD process can be formed without damage to the thin film. They can be laser cut, waterjet cut, and sheared.

There are three main PVD processes:

1. Vacuum evaporation
2. Sputter deposition
3. Arc vapor deposition

The most common processes used on sheet are vacuum evaporation and sputter deposition. Arc vapor deposition is used more commonly on smaller parts, such as hardware and faucets.

To create a PVD coating, a vacuum is created and argon along with nitrogen are delivered to the chamber. A high-current source is applied, creating a plasma state in the argon-filled chamber (Figure 3.38).

In all of the PVD processes there is a target material, such as titanium, zirconium, chromium, and other metals. The target material is made a cathode and when the plasma state develops in

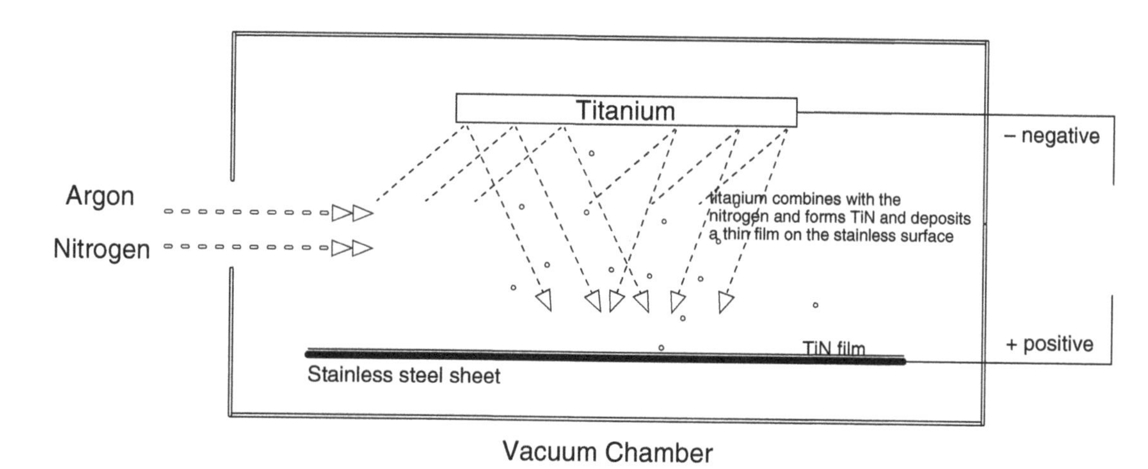

FIGURE 3.38 PVD sputter deposition process.

the vacuum chamber, argon atoms become positive and highly charged. They collide against the material of the cathode, releasing atoms in ionized vapor form. The substrate (in this case stainless steel) is made anodic, and the negatively charged ions combine with the nitrogen and deposit as nitrides on the surface of the stainless steel, forming a thin, unbroken film.

New techniques have been developed that involve magnetrons to concentrate the plasma near the target material to be bombarded. This further increases energy use but speeds up the process. The process involves an extensive array of dedicated equipment to prepare the sheet and apply the finish. For a successful coating application, the need for establishing a quality process is critical. The finishes produced are very durable but can fail after exposure if critical controls are not followed (Figure 3.39).

Vacuum evaporation is similar. An electron beam targets a material in a strong vacuum. The atoms precipitate and coat everything in the chamber in line of sight of the bombarded material.

The thin coatings used on stainless steel are combinations of nitrides, carbides, and oxides, depending on the gas added to the argon in the vacuum chamber. As the targeted material is ionized, different gases are added at controlled rates to combine with the target material and carry out deposition on the surface of the stainless steel. Temperature in the chamber will determine the density and structure of the coating created. This in turn determines color. Adjusting the rates and concentrations of the various compounds can create color tones. Certain colors and materials are more commonly used than others.

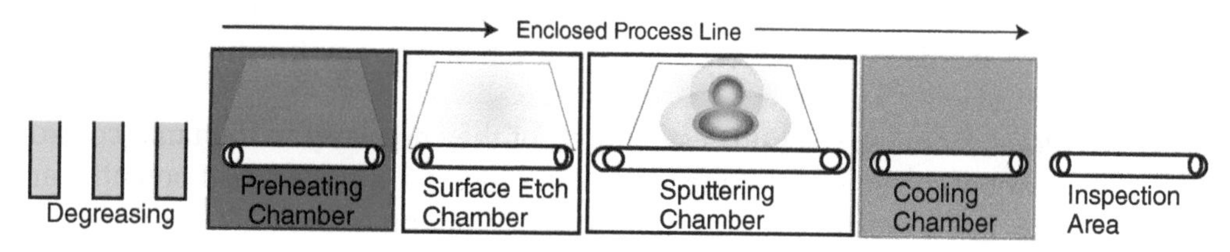

FIGURE 3.39 PVD sputter deposition process line.

Titanium nitride, the gold color used to coat and make drill bits harder, is a coating commonly applied to stainless that produces a beautiful gold color. Titanium carbide is used to create a deep black color, while zirconium nitride produces a range from light golden to dark red.

Compounds Commonly Used in the PVD Process to Generate Color

Titanium nitride
Titanium carbide
Chromium nitride
Titanium-aluminum nitride
Zirconium nitride

Tubes, bars, sheet, and other forms can be coated using PVD. Finishes on the stainless steel surface, including glass bead, mirror, Angel Hair, Vibration, and other surface variations, can be coated. The combinations of color and finish can produce striking surface colorations. Unlike the Inco coloring process, these finishes do possess color. Their thinness allows for some interaction with light interference, as light will pass through them and will reflect off of the surface.

Colors that can be developed with PVD on stainless steel include:

Gold
Black
Bronze
Blue
Golden silver (often referred to as nickel silver)
Red/purple

Variations within these basic colors can also be achieved to produce copper-like colors and reddish bronzes. The surfaces can be masked to produce decorative highlights, with the natural color of stainless showing through. Sheets of PVD can be embossed, formed, and shaped. They can be laser cut or waterjet cut. Post-welding will damage the finish, though, and limitations in chamber size will restrict how large a sheet or fabrication can be coated with the process. Plates can also be coated, but again, how robust the vacuum chamber is and the handling system of the coating company may limit the size and thickness of plates.

As with all processes that add coatings intended for aesthetic and performance applications, there are tight parameters the finishing companies must adhere to. In all processes, clean surfaces are demanded. Clean stainless steel sheets are a requirement for any finishing process, whether mechanical or chemical.

In the Inco colored stainless steel process, control of the chemistry in the tank, control of the rinse tank, and control of the emersion time as it relates to the films' electrical potential are a few steps that are critical for successful coloring. For the PVD coating to be successful, careful control of the temperature and makeup of the introduced gases are critical to the process. There have been instances in successful projects for both the Inco and the PVD processes in which the project deteriorated prematurely due to film breakdown once exposure to ultraviolet radiation and ambient conditions was experienced. The phenomena of film breakdown are not fully understood. It does not happen right away. Panels can be fabricated, delivered, and installed, and months later changes in the surface color will develop, meaning that something external has had an effect on the surface. In the Inco process, color interference films in the thicker range, such as the greens and reds, have experienced rare premature deterioration. In the PVD process, failure to control the temperature as the plasma is initiated has led to rare instances of color transformation. In these occurrences exposure to ultraviolet light played a role, as hidden returns and concealed and shaded areas remained unaffected.

Cleaning of Colored Stainless Steels

Colored stainless steels, like natural stainless steels, are susceptible to damage from deicing salts and iron contamination. Spots will appear as the chloride ion attacks the surface. The surface must be cleaned. The stains can be removed without damage to the color if they are addressed quickly. Use of abrasives will damage the color. Consider detergents and mild solvents; that is, only cleaners that will not dissolve the chromium oxide layer. Diluted phosphoric and citric acid, if quickly applied and removed, will work. Electropolishing and nitric and hydrofluoric acids can remove the chromium layer. Deicing salts and chlorides have been shown to attack the surface and even alter the color of green and red stainless steels. It is advisable to clean the surface regularly with fresh water, mild detergent, and deionized water.

ELECTROPOLISHING: BRIGHTENING THE FINISH

Stainless steel is well suited for the process known as electropolishing. Electropolishing is an electrochemical process similar to electroplating, but instead of depositing metallic ions onto a metal surface, metal ions are removed. Electropolishing will increase the chromium to iron ratio near the surface, which benefits the corrosion resistance of stainless steel. Do not use electropolishing on interference colored stainless steels. Electropolishing will remove a thin layer of oxide and this will remove the color.

Sometimes referred to as "electrocleaning" or "chemical polishing," this process differs from mechanical methods of polishing and buffing stainless steel. In mechanical polishing, stresses can develop as metal is removed and as the microscopic surface of the metal is smeared and torn from the abrasive action. Heat energy is also induced into the metal as it is mechanically polished. This leaves a thin stressed layer on the surface. Electropolishing removes metal from the surface by dissolving it electrochemically. Heat is not added and stresses in the surface do not develop.

The electropolishing process can be performed on mechanically polished surfaces to remove some of the disturbed layer of metal left after the mechanical abrasion.

The advantage of electropolishing is that the resulting surface does not have any loose, sharp surface anomalies that can come off later or trap contaminates. Edges of microscratches are rounded as the points are dissolved in the electropolishing process.

Corrosion resistance is improved by removing areas that can trap contaminates and weak layers of metal are dissolved.

One minor drawback to electropolishing is the way surface defects in the stainless steel – such as surface pits, voids in the surface, seams, and metallurgical defects – are revealed. Unlike mechanical polishing, which can hide defects by interfering with the reflection, electropolishing will reveal defects.

In the electropolishing process, the stainless steel part or sheet is made the anode in an acidic electrolyte bath. Direct current is applied, the stainless is connected to the positive electrode, and a copper, lead, or stainless steel plate is attached to the negative pole. As a low current is applied, metallic ions are removed from the stainless steel anode. High points projecting out from the surface become areas of higher current density relative to lower regions and are dissolved into the acid solution as ions. Essentially two layers develop on a metal surface undergoing electropolishing. The first layer is composed of a saturated or supersaturated product of dissolved metal ions and electrolytes. This layer is usually shrouded with a second, gaseous layer typically composed of oxygen gas discharged from the electrolyte during the process. These layers effectively blanket the metal and follow the macrosurface of the metal in such a way that depressions on the surface are covered with a thicker layer while projections have a thinner layer. This saturated film provides resistance to the current. Thus, the current is concentrated around the projections from the surface where the layer is thinner. This concentration at the projections dissolves the metal more rapidly, leading to a microleveling of the metals surface.

Burrs, undercuts, and free iron on the surface are removed, leaving a smoother, more corrosive-resistant surface. It is important to note that the surface achieved from electropolishing directly corresponds to the surface prior to electropolishing: it will smooth out high points and edges, but it will not remove scratches, pits, scale remnants, cutting, and forging marks, or other surface maladies that are greater than 0.001 in. (25.4 μm). Such surface maladies may be accentuated in the electropolishing process.

Typical electropolishing baths are heated solutions of acids.

Common Electropolishing Baths

Phosphoric acid

Phosphoric acid and sulfuric acid

Glycolic acid and sulfuric acid

Citric acid and sulfuric acid

Depending on the desired results, the process only takes a few minutes. It is a rapid and more thorough process compared to buffing or hand finishing. Corners and edges are polished far more easily than with mechanical techniques. It does require a setup similar to a plating operation, with necessary rinse tanks and means of handling and disposing of acids and metal sludge. Therefore, it is removed from the normal processes of metal fabrication firms and requires a specialized processing operation.

To further benefit the surface created from electropolishing, the iron and nickel near the surface are dissolved more readily than the chromium. This in turn benefits corrosion aspects further by leaving a chromium-enriched surface behind. Free iron, oxides, and areas where corrosion can initiate are removed.

The resulting surface is always smoother and brighter than before electropolishing. The microdissolution of surface irregularities eliminates conditions that cause light scattering. The appearance can be described as a silvery white luster, much different than that obtained from mechanical techniques. The process does not change the dimensions of the part, only removes microscopic features from the edges and surface.

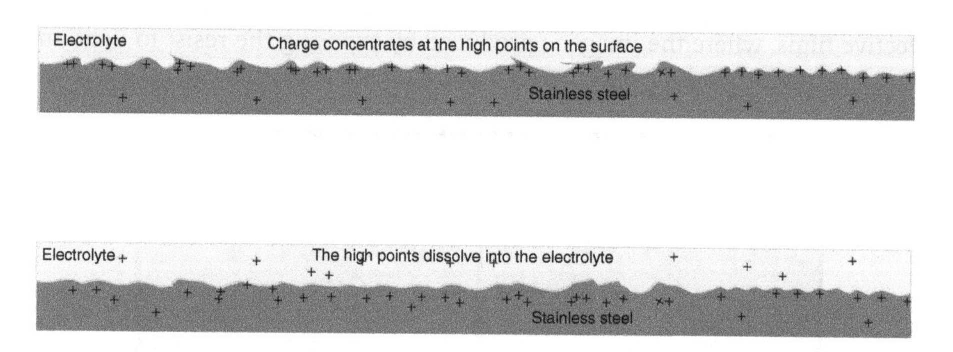

Electropolishing can be performed on any shape that can fit in the chemical bath. Perforated plates and sheets formed parts and castings can be electropolished. As in electroplating, there are nuances in setting up the tank to deliver the current density effectively to treat the entire surface. Also as in electroplating, a very clean surface, free of oils, grease, and excessive oxides, is a requirement for good results. Once electropolished, the part needs to be thoroughly rinsed to remove the electrolytes and dried.

Electropolishing neither induces stress into the surface nor has the potential to embed microscopic particles of any kind. Mechanical polishing and subsequent buffing operations effectively cold work the metal and have the potential for embedding foreign substances into the surface, which can later become corrosion initiation points if the circumstances present themselves. Electropolishing of perforated surfaces refines the edges of the perforations, removes any microscopic burrs, and removes any minute steel particles that may have been extracted from the steel punch and dies.

Satin finish surfaces, when electropolished, become smoother and brighter. They will not match other satin surfaces that have not been electropolished. The luster brightens as well due to the smoothness and the chromium enrichment.

ETCHING

Stainless steel surfaces can be etched using various acid solutions to dissolve the metal. Resists can be applied to create designs on the surface. Resists are adhered to the surface or painted onto the surface. The acids are allowed to attack the exposed metal surface and dissolve it. Usually the surface is facing down so the dissolved metal will fall into the solution and exposes more metal. There are several proprietary solutions used to etch designs into stainless steel surfaces. Milder systems using electropolishing equipment will etch into the surface.

Acids that are used on stainless steel are very powerful and should be handled with extreme care. A ferric chloride solution, which becomes hydrochloric acid when mixed with water, is a common etchant. Sulfuric acid and nitric acid will also etch stainless steel.

The use of photoresists can produce intricate detailed art on a surface. Resists can also be hand cut (Figure 3.40).

There are other means of producing designs on the surface of stainless steels. One is using photoresist protective films, where the image is produced by exposing the resist to light and allowing

FIGURE 3.40 Acid etched stainless steel.

FIGURE 3.41 Photoresist and color removed from mirror black surface in an art prototype for artist Jan Hendrix by Rimex Metals.

the decorative feature to be exposed for etching or hand-cut protective films. The protected stainless steel sheet is passed through a glass bead blasting machine and the glass bead texture is selectively induced in the regions that are exposed. The photoresist allows for a level of production on multiple sheets of stainless steel. This can also be used to selectively color stainless steel. The color goes into the stainless steel where the resist is not (Figure 3.41).

Laser etching using CNC can "burn" an image into the surface of stainless steel. Very fine detail can be produced. The image will be a contrasting black color on the silver background. This is not often used on large plates, and there will be some corrosion sensitivity on the darkened surface. The detail is similar to what can be obtained in a photographic image (Figure 3.42).

A waterjet can also etch designs into a stainless steel surface. Essentially the waterjet carves the design into the surface by penetrating to a depth of only a few percent of the total thickness. Waterjets can achieve good detail, but not as fine as other etching processes. The cost is lower than laser, acid, and photoresist color removal processes (Figure 3.43).

BLUING

Bluing or darkening of a stainless steel surface to give it a feel more like steel and to reduce the shine is more an artistic process than an architectural one. Selected small areas, furniture, and occasionally panels have been darkened using various bluing chemistry.

FIGURE 3.42 Laser etching of stainless steel at the Greater North Charleston Naval Memorial, designed by BNIM.

There are several proprietary mixtures in use that have differing degrees of success. There are methods of achieving blacks and dark blues. Some are hot processes and some are cold processes. It is important to note, however, that the finish is often mottled and uneven with dark and light zones.

To produce these finishes, the protective chromium oxide on the surface has to be removed for the base metal to be attacked by the chemicals. There are several ways to do this. You can physically remove it with mechanical means just prior to applying the chemistry, or you can chemically remove it.

The chemical process is not the easiest and does have some limitations, both in terms of the environment and performance. The rinse and disposal requires special handling and care as well as safety equipment for the persons applying the chemicals.

Additionally, the process may affect long-term performance expectations for the stainless steel because of the destruction and chemical alterations of the chromium oxide film.

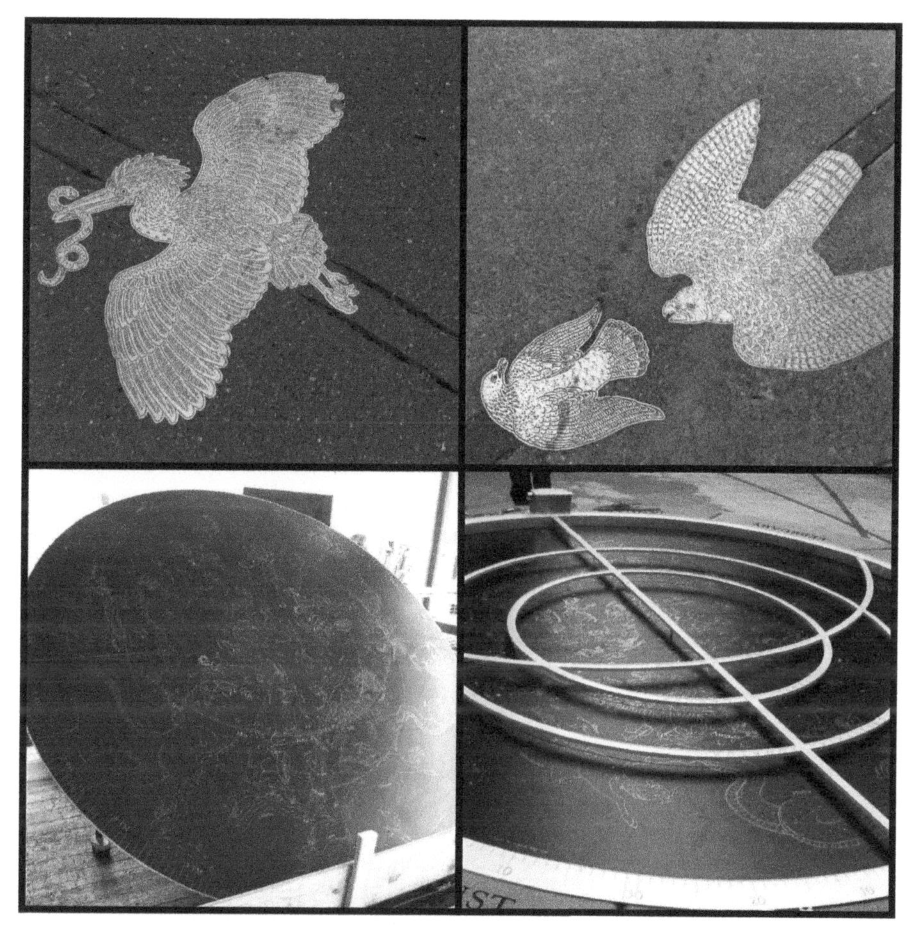

FIGURE 3.43 Waterjet etching of stainless steel birds and the blue stainless steel *Star Disk* in *Celestial Flyways*, Kansas City, Missouri, designed by artist Laura DeAngelis.

The numerous finishing processes and combination of finishing processes available to the artist and designer are staggering. Listed here are mainly the commercially developed processes. The ability to change and alter the finish of stainless steel is one of the most fascinating attributes of the metal. Few other materials known to mankind offer as many choices. The past 40 years have added an enormous number of finish possibilities to stainless steel. What once was defined as a series of numbered finishes has expanded to commercially viable, unique surfaces finely tuned by the artist and architect. The next 40 years should offer even more. Understanding how these choices are applied and how they will perform over time and exposure is critical.

CHAPTER 4

Expectations of the Stainless Steel Surface

Surfaces define the shape of this world, light allows us to see them

Herbert George, *The Elements of Sculpture*

Of all the metals used in art and architecture, stainless steel can be one of the most challenging. The attributes of stainless steel that provide the beneficial characteristics we seek in the metal can sometimes work against successful outcomes from an aesthetic viewpoint (Table 4.1). The resilience and strength of stainless steel come with internal stresses within the sheet or plate. The surface finish and the reflective nature of the finish amplify minor surface deviations these internal stresses may create. Corrosion resistance, the stalwart characteristic of the metal, can be influenced by the finish applied to the metal and the methods used to apply the finish.

There are several variables that influence successful outcomes when using stainless steel as an aesthetic cladding material. In this chapter these variables will be examined from an appearance perspective as to their influence on the flatness of a finish panel, the surface hue and tone of an overall surface, and the corrosion behavior of the finish.

REFLECTIVITY: BRIGHTNESS AND INTENSITY

The beauty of the stainless steel surface is derived from the way it reflects light. The inherent luster the metal possesses can give the feeling that light is being generated from the metal itself. Diffuse surfaces scatter the reflected light and the true color of the metal is apparent to the viewer. The surfaces can reflect the subtle glow of their surroundings, appearing blue when the sky is blue or reflecting the glow of the colors of a sunset.

TABLE 4.1 Variables that can have an effect on an aesthetic outcome.

Internal variables

- Alloying constituents
- Annealing atmosphere
- Casting heat
- Grain size

External variables

- Surroundings
- Angle and distance in viewpoint
- Effects of fabrication
- Effects of storage, handling, and installation
- Thermal changes

Material surface variables

- Mill finish quality
- Mechanical finish
- Directionality

Certain finishes will reflect more of the light and can produce the disturbing condition of glare when bright sunlight hits the surface. Often one of the first questions people raise is, Will the surface be too glaring in bright light? Will the reflection interfere with airplanes or automobiles? These are questions to be considered, and they have a lot to do with the choice of finish, color, and geometric position of the surface (Figure 4.1).

There are two categories of reflection: diffuse and specular. Most of the materials that make up our environment have a diffuse reflectivity because their surfaces are not smooth but porous and rough. Water can be specular in reflectivity, as can glass, but stone, concrete, and most metal surfaces (particularly those with an oxide) will scatter light, creating a diffuse reflectivity. Stainless steels can have surface finishes that are specular, diffuse, or a combination of both. Stainless steels can be given colors that deconstruct a portion of the reflected wavelength in such a way the reflective glare is altered, similar to tinted glass.

The Jay Pritzker Pavilion at Millennium Park in Chicago, Illinois was designed by Frank Gehry and is clad in stainless steel with an Angel Hair finish. It faces the south, so it receives direct sunlight. There is no offensive glare from this surface because it effectively scatters the reflected light (Figure 4.2).

It is a combination of the diffuse Angel Hair finish and the convex shape of the curving geometric forms that make up the surface. Not far from the pavilion is Burberry's flagship store

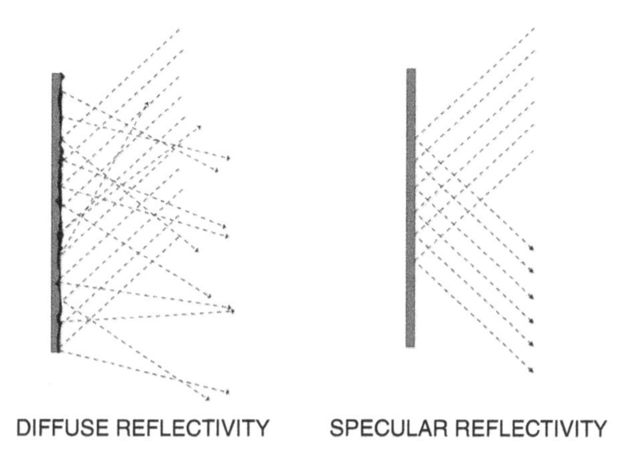

FIGURE 4.1 The two forms of reflection.

FIGURE 4.2 The Jay Pritzker Pavilion at Millennium Park in Chicago, designed by Frank Gehry.

FIGURE 4.3 Burberry's flagship store in Chicago.

in Chicago. The surface is a Super Mirror black stainless steel made up of numerous flat panels. The black is induced by using the physical vapor deposition (PVD) process with titanium carbide. The surface is highly reflective, but the reflected light wave lacks the intensity contained in the yellow or golden portion of the spectrum (Figure 4.3).

When color is introduced into the stainless steel surface via the techniques of light interference or PVD, it can effectively alter the glare generated from specular reflectivity. What occurs is a partial cancellation of part of the visible spectrum that is reflecting back to the viewer. In Tampa, Florida, the Museum of Science and Industry designed by Antoine Predock has a large spherical form clad in mirror blue interference stainless steel. Here again, the shape and the color of the stainless steel effectively reduce offensive glare. The convex shape avoids concentrating the light and the reflected light has some interference-induced destructive cancellation of the light wave to give the sphere a blue color (Figure 4.4).

In Cleveland, Ohio, there are several jewels that adorn a portion of the city. Each of these has various finishes and colors and do not create offensive reflective glare. The Museum of Contemporary Art designed by Farshid Moussavi is wrapped in flat planes of mirror dark blue stainless steel. The angular façade wraps the building at all elevations. The specular reflectivity from this surface is akin to blue sapphire and does not create offensive glare to traffic or the people living in nearby residences (Figure 4.5).

FIGURE 4.4 Museum of Science and Industry in Tampa, Florida, designed by Antoine Predock.

FIGURE 4.5 Museum of Contemporary Art Cleveland, designed by Farshid Moussavi.

Not far from the museum is Case Western University. Adorning the roof of the Peter B. Lewis Weatherhead School of Management is a remarkable surface designed by Frank Gehry. The undulating surface is composed of thin stamped shingles of natural-color stainless steel with a directional No. 4 finish. This surface appears at times as if the light is emanating from the curving shape. In bright sunlight there are areas that are intensely reflective when the angle of incidence is near parallel to the grain. However, the shape of the surface and the texture of the shingles produce a shimmering effect rather than a glaring condition (Figure 4.6).

A short distance from these projects is the Cleveland Museum of Natural History and the Shafran Planetarium. The surface of the planetarium entrance is clad in an Angel Hair stainless steel colored with a PVD coating. This surface is a stunning reddish bronze titanium-based PVD coating that allows the reflective nature of the stainless steel to gleam with the color induced by the thin titanium compound layer (Figure 4.7).

The directional finishes – No. 3, No. 4, Hairline, and the custom finishes, such as Cross-Fire – have a polarized diffuse reflectivity. They will reflect partially specular in one direction and

FIGURE 4.6 Weatherhead School of Management, designed by Frank Gehry.

FIGURE 4.7 Shafran Planetarium, designed by Westlake Reed Leskosky.

partially diffuse in the orthogonal direction. At some viewing angles, where the angle of incidence strikes along the parallel lines, there can be an intensive reflectivity. The intensity is reduced by the diffuse reflective aspect of the reflection that also occurs. The Hairline finish is much finer than the No. 3 and No. 4 finishes. The graining introduced to the surface can demonstrate a specular reflectivity significantly more intense than the deeper, coarser finishes of the No. 3 and No. 4 finishes. Usually this finish is used on interior surfaces, where the concern for glare is less of an issue (Figure 4.8).

FIGURE 4.8 Contemporary Jewish Museum in San Francisco, designed by Daniel Libeskind.

The glass bead finish is a diffuse texture induced onto the surface of stainless steel. The glass bead surface has a higher gloss than the Angel Hair finish unless it is applied over a directional satin finish or a 2D. The glass bead finish creates small craters that effectively scatter the light. When applied over a directional finish, this creates a flat, matte diffuse tone.

At the 9/11 Memorial & Museum in New York, designed by Snøhetta Architects, a multiple layering of finishes creates the magnificent façade. There are directional finishes similar to a No. 4 with overblasting of glass bead stripes to further mute the reflectivity. The façade is gleaming and lustrous without generating intensive glare on its surroundings. The directional, more reflective No. 4 surface is broken by the striping created by selectively bead blasting, so that what glare is generated is reduced. There are angles where the sun reflects directly off the directional finish bands, but not unlike the sea, the reflectivity has a limited intensity restricted by the matte finish bands (Figure 4.9).

Reflective glare from metal surfaces is a design constraint that should be evaluated. There are a number of technical specifications, but creating a prototype of the surface and viewing it at different angles and at different times is an essential way of arriving at validation. Gloss meter readings of

FIGURE 4.9 National September 11 Memorial Museum, designed by Snøhetta Architects.

the reflective levels of the metal and luminosity readings act as guides to close in on the level of reflectivity that will be experienced. But a prototype of the surface is the best way to determine whether a reflection will be offensive or not. Often there are subtle ways to adjust the reflective nature of the surface that will benefit the design and restrict the uncomfortable glare from becoming an issue.

Measuring gloss is a method that can be used early in design. Gloss measurements of diffuse surfaces can direct the design to a texture that has minimal glare. It does not take into consideration color and how color will also have an effect on the discomfort or lack of discomfort induced by the reflectivity. Gloss is not directly related to the amount of energy of light reflected from a surface. The amount of energy of light increases as the angle of illumination increases. For low-gloss, matte surfaces, the angle of incidence of a beam of light will not be the same as the geometric angle created by the light beam arriving at the surface. The light beam will be reflected at the specular angles created by all the individual surface irregularities. This makes it a specular reflection.

Certain code requirements are establishing daylight reflectance levels to limit the occurrence of glare or bright reflectance off surfaces. Aviation entities as well are concerned about reflective surfaces and their effect on pilots.

The term “daylight reflectance,” in the context of a materials surface, refers to the total sum of both diffuse and specular reflectance from that surface. In reality, it is the specular reflection that is the concern. The diffuse reflectivity does not present an issue for surrounding buildings and people, as shown by the Pritzker Pavilion or other stainless steel surfaces with a diffuse finish.

Dallas/Fort Worth Airport’s Terminal D has a stainless steel roof that covers several thousand square meters. Its finish is a diffuse, rolled, coined surface finish over a mill 2D surface. There is no glare from this matte surface even in the bright sunlight. Similar to this finish is the one on the walls of the Tacoma Art Museum. These flat walls of stainless steel show no specular reflection (Figure 4.10).

FIGURE 4.10 DFW Roof and Tacoma Art Museum with roll applied coined finish over a 2D mill finish.

These natural finish stainless steel surfaces have low gloss readings, below that of the Angel Hair finish on the Pritzker Pavilion. The Ronald Reagan Washington National Airport has a roof covering of stainless steel. The finish on this surface is a mill finish similar to a 2D surface. The reflectivity is diffuse and poses no issues with glare.

Figure 4.11 shows the curving stainless steel facade designed by Rojkind Architects for the Liverpool–Interlomas department store in Mexico City. The surface is a combination of Angel Hair and a rolled on coined finish. In bright sunlight the surface exhibits some specular levels along with the diffuse scattering surface. The effect is more of gleaming elegance than offensive glare. It is a statement of modern architecture.

The Archway, located directly over I-80 near Kearney, Nebraska, is clad in stainless steel corrugated panels with interference colors to mimic the Nebraska sunset. The designer wanted a blend of colors across the surface. The concern was the glare that would be produced from the reflection off the surface. Initially the surface was to be glass bead blasted stainless steel with interference colors induced after bead blasting. Mock-ups of the various colors found the reflection from the

FIGURE 4.11 Liverpool–Interlomas department store, designed by Rojkind Architects.

FIGURE 4.12 The Archway, Kearney, Nebraska.

yellow/golden colored stainless steel created a condition of glare that might pose discomfort under certain angles of reflectivity. The yellow/golden color was desired, but a method to further induce an additional level of diffuse reflectivity was examined. The approach used took the surface of the stainless steel and added an embossed texture on top of the glass bead. This inexpensive adjustment enabled the golden color to be used and achieved the necessary diffuse reflectivity to eliminate potential glare (Figure 4.12).

Tests performed to determine the levels of reflectance of various metal finishes show the total reflectance as being at moderate levels but the specular reflectance being low (Table 4.2). The readings in the table are samples, but they give an indication of the level of reflectance from various finishes and correlate with gloss readings. It is interesting to note that the readings of specular reflectivity are higher for the non-directional Angel Hair finish than for the glass bead. Yet Angel Hair has a lower gloss reading (Figure 4.13).

TABLE 4.2 Total reflectance readings for selected non-directional metal surfaces.

Metal and finish	Total diffused light reflected (%)	Total specular light reflected (%)	Total light reflected (%)
Aluminum with matte paint	47.72	1.09	48.81
Stainless steel: 2B mill finish	43.65	8.58	52.23
Stainless steel: glass bead	45.14	2.15	47.29
Stainless steel: Angel Hair finish	40.66	4.44	45.10

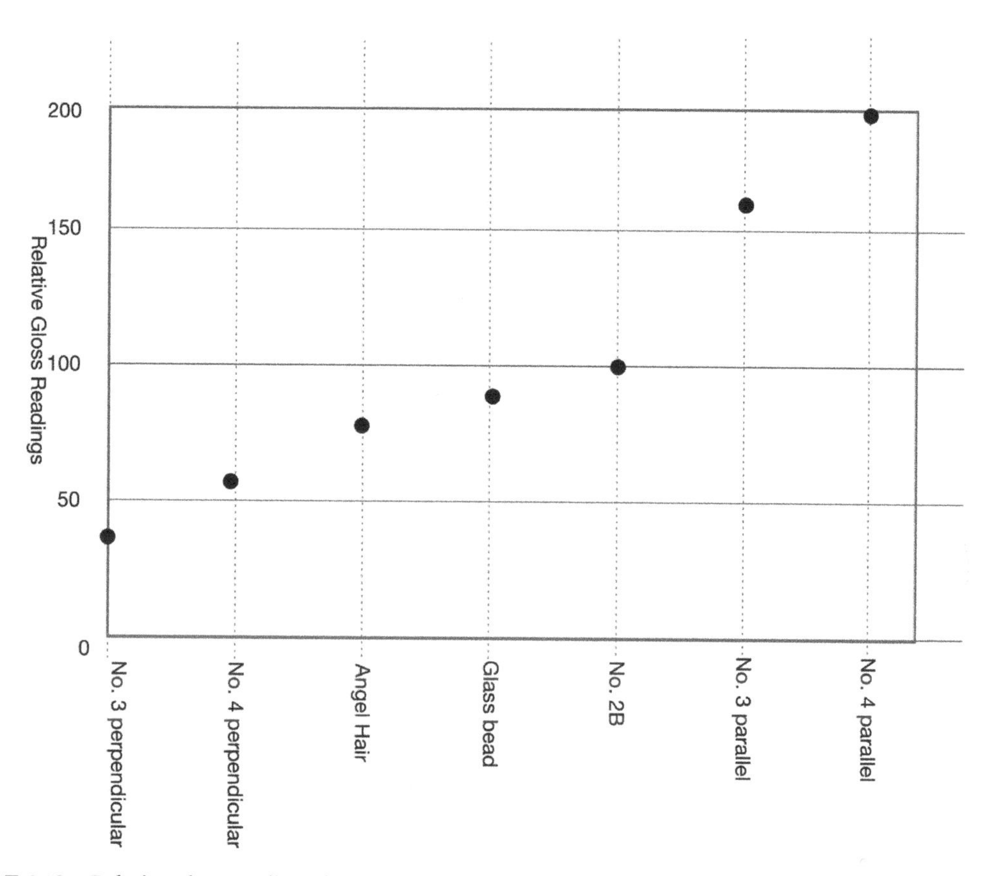

FIGURE 4.13 Relative gloss readings for diffuse surfaces.

SURFACE FINISH AND REFLECTIVITY

Surfaces in art and architecture are often composed of an extensive array of metal elements. This array could be made up of overlapping or butted flat or formed panels made from sheet or plates. The surface could consist of an extensive flat or curved monolithic form wrapped in a shimmering skin. Or the surface could be composed of facets that intersect with open seams or welded joints. Because these surfaces are made up of layered or interlocked components that create an extensive, monolithic enclosure, subtle differences in appearance will be readily apparent, as the slightest change in plane enhances the reflective contrast between elements.

Many of the finishes applied to stainless steels have different levels of diffuse and specular reflectivity. The diffuse reflection scatters light and enhances the metallic color one sees. The specular reflection can affect the color, but also can induce "hot spots" where the reflection is intense (Figure 4.14).

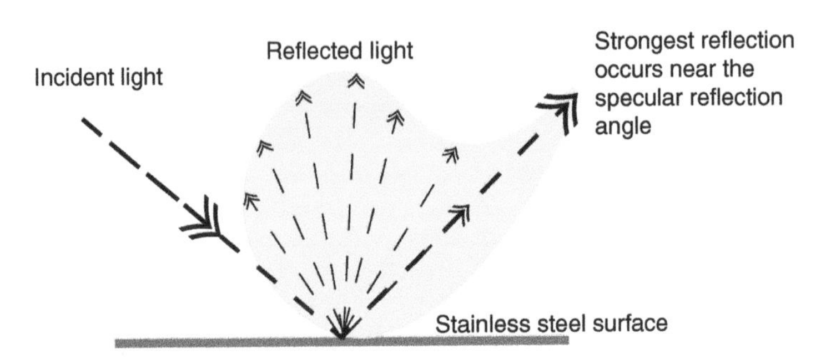

FIGURE 4.14 Diagram of reflection, both diffuse and specular.

These conditions also induce levels of contrast where minor differences on the surface can appear greater than they actually are. As you move away from the surface your eye will perceive light and dark contrasting regions, not unlike the way a body of water appears as wind or light waves pass over the surface: up close they appear smooth, but at a distance minor differences begin to appear. Figure 4.15 shows an Angel Hair surface viewed at close proximity, at 100 m out, and at 1000 m out. Up close (the upper left image), the surface appeared consistent and smooth. As you move away from the surface, distortions began to appear, first visible at approximately 100 meters (see upper right image). At a great distance, distortions are more apparent, as seen in the lower image taken from an office tower several blocks away. After several months, much of this distortion in the surface of the building was no longer apparent because the building surface and the building structure had undergone thermal changes. The expansion and contraction of the different components eventually reached equilibrium. The skin is composed of many small elements that move due to thermal changes, whereas the building itself is composed of larger elements that move at different rates. After a couple of temperature cycles, the surfaces change and a different alignment occurs.

Such is the nature of reflective surfaces. Stainless steels are one of the most reflective materials utilized in architecture and art. Stainless steel captures the light and sends it back to the viewer. Subtle variations are therefore more apparent. Often these variations are contingent on the light source, angle of view, and distance from the surface. While the surface can look smooth and consistent in one light condition, in other lighting and viewing circumstances the same surface can take on visually different appearances.

The color and tone of stainless steel will vary with atmospheric conditions, alloy composition, heat treatment processes at the Mill, and surface finish characteristics. Photometric studies[1] on the evaluation of stainless steels in the atmosphere show the chromaticity of stainless steel surfaces varies primarily in the yellow hues, while little variation is apparent in the red and green hues. It has been shown that the chromaticity of stainless steels increases in yellow hues and decreases in lightness of color with exposure to a moist, humid environment. The cause of this is unclear. Most likely it is due to the absorption and reemittance of light from the iron and nickel components in the

[1] Kearns, J.R., Johnson, M.J., and Pavlik, P.J. (1988). The corrosion of stainless steels in the atmosphere. In: *Degradation of Metals in the Atmosphere* (ed. S.W. Dean and T.S. Lee), 35–51. American Society for Testing and Materials.

FIGURE 4.15 Distortions begin to appear as you view from increasing distances.

alloy that are at or near the surface. Some of this can be apparent under changing light conditions, as when the sun moves across a surface as shown in Figure 4.16.

Metals in a natural uncoated form will vary in appearance from one component to another under differing light energy levels. Whether alloys, such as stainless steel and aluminum, or commercially pure copper or zinc, metals will not be as consistent in color and tone as a painted or resin-coated surface. Stainless steels in particular will display inconsistencies in different light energies and at

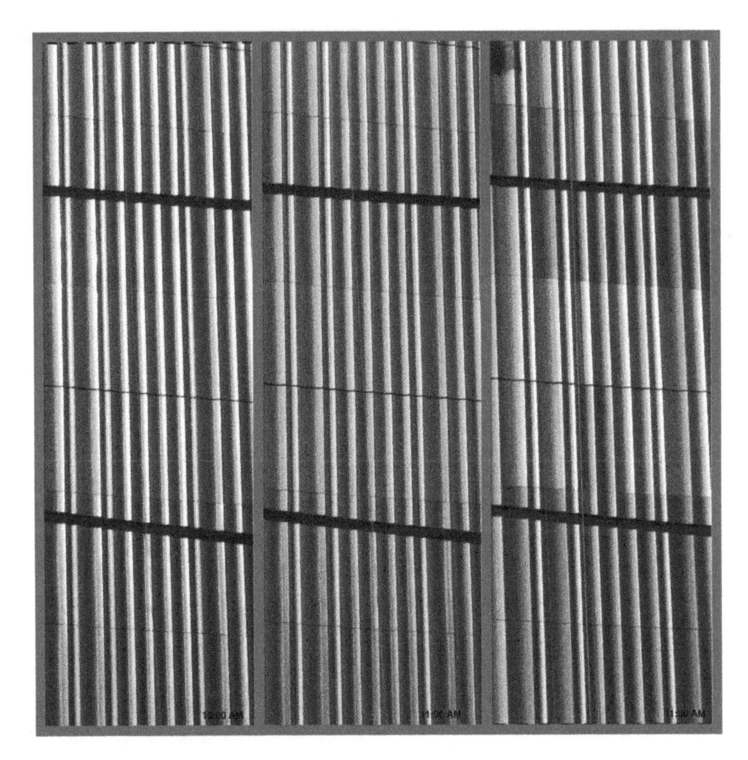

FIGURE 4.16 Color differences under different lighting conditions.

different focal points more readily than most other metals because of their reflective luster. This phenomenon is most pronounced in light-scattering, diffuse surfaces than in highly reflective, polished surfaces. The reason for this is that on light-scattering diffuse surfaces, the scattered light particles will exhibit selective absorption in different regions of the visible spectrum and impart a color tone to the object. Energy in light is a function of wavelength. The lower wavelengths (the blue end of the light spectrum) carry higher energy levels (Figure 4.17). On cloudy, overcast days, this portion of the light wave is muted, which can cause different color tones to appear on the stainless steel surface. On bright days, the full spectrum is at play and the metal surface appears to match. This condition can also show itself at night, when artificial light illuminates the surface. Artificial light sources normally do not emit the full spectrum.

The stainless steel surface appears monolithic in color and tone in certain light energy levels during the day, while as the light energy changes, tonal differences manifest in one element in contrast to adjoining elements. Whereas these elements matched at one point, they no longer match as the light condition and light energy levels change. The appearance of the elements can sometimes flip, with one element appearing darker under certain conditions and lighter under other conditions. On some stainless steel surfaces the finish and reflective color may look very consistent in bright sunlight, only to show patchy, irregular coloration on cloudy days, or vice versa.

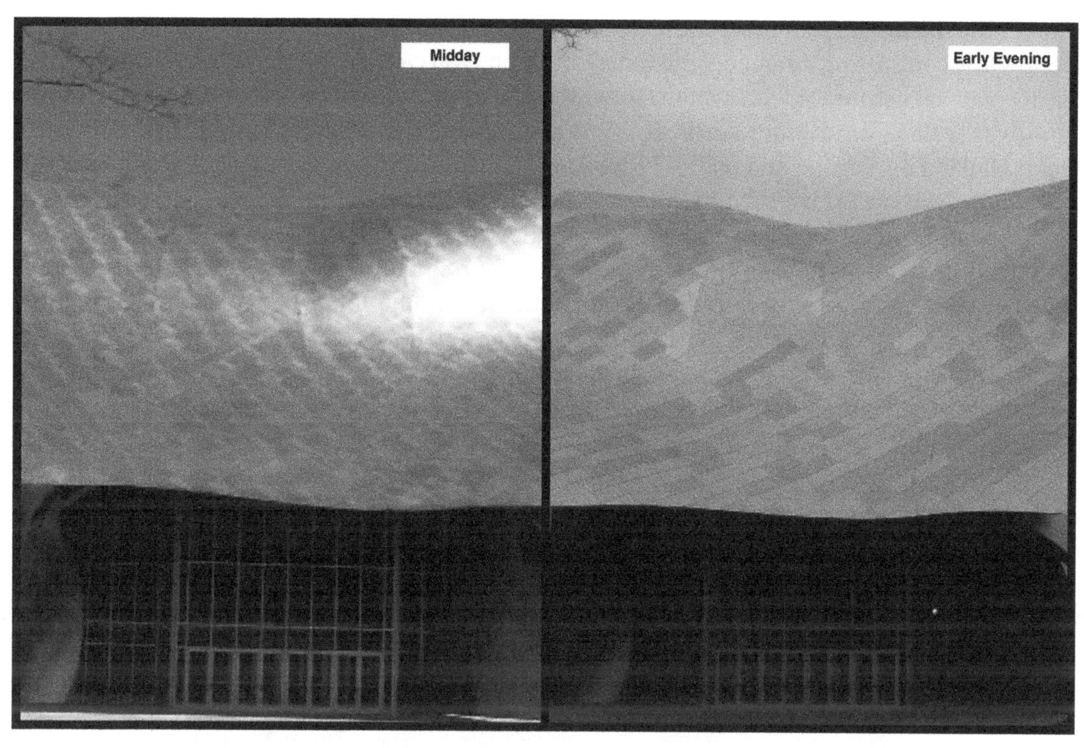

FIGURE 4.17 Stainless steel reflective differences at midday and early evening light on the Fisher Center at Bard College, designed by Frank Gehry.

Figure 4.16 shows the same surface on the same day 90 minutes apart. As the sun moved higher in the sky a color shift appeared, and as the day progressed the color shift disappeared.

This behavior is not due to the phenomena known as metamerism[2] that paints sometimes exhibit, but is often attributed to wear on the equipment that was used to produce the diffuse surface of the stainless steel. This can be the case, but isn't always. When the polishing belts or discs are worn, or different grit is used, the finish will be different in different areas, but this contrast will be pronounced in many lighting conditions. The phenomena referred to here is not easily measurable. It is difficult to correlate the microscopic surface with the observed finish behavior. The differences are not easily correlated with gloss meter or colorimeter measurements. Differences in color and tone are not discernable when viewed from close distances either, but are only manifest when viewed from a distance or at a grazing angle of light.

This phenomenon is difficult to prevent or predict. On large surfaces made of multiple elements and in particular stainless steel plates the phenomena are common. The thinner the sheet metal,

[2]Metamerism is related more to color and the differences in distribution of energy from different light sources. A color can look one way under fluorescent lighting and another when viewed in natural light.

the lower the likelihood is of this condition of contrasting elements. As the thickness of sheet metal increases, the contrasting tones become more common. You can see this in large surfaces of stainless steel: in plate structures, such as the Gateway Arch in St. Louis and the Air Force Memorial in Virginia, as well as on thinner surfaces, such as the stainless steel cladding on the Walt Disney Concert Hall in Los Angeles and on the Fisher Center at Bard College (Figure 4.17). Thin sheets used on the BP Pedestrian Bridge at Millennium Park show this condition (Figure 4.18). Some finishes, such as the linen finish, are composed of a mill-created surface and a mechanically imparted coining. The shape is induced under significant pressure and is produced on the face surface only. The finish can be a beautiful surface. Because of the way it is made, the top/outermost surface retains the nature of the mill finish because it is pressed into an opening on a roll surface. The opening is deeper than the pattern is high, so the metal does not bottom out into the roll. The metal around the edges receives the texture on the balance of the roll. This makes the tops of the protruding surface more specular than the rest of the metal, which has a diffuse texture induced into it. There is a so-called sluff line introduced into the pattern, which gives it its distinctive linen look. This "line" is actually not a line, but instead is where the protruding bumps are placed closer together to break up the pattern. Figure 4.19 shows a linen finish with the line running across the width of the sheet. It serves no purpose other than to add detail to the metal. A close-up of the design is shown in the image in the lower right-hand corner.

FIGURE 4.18 Contrasting tones on thin stainless shingles.

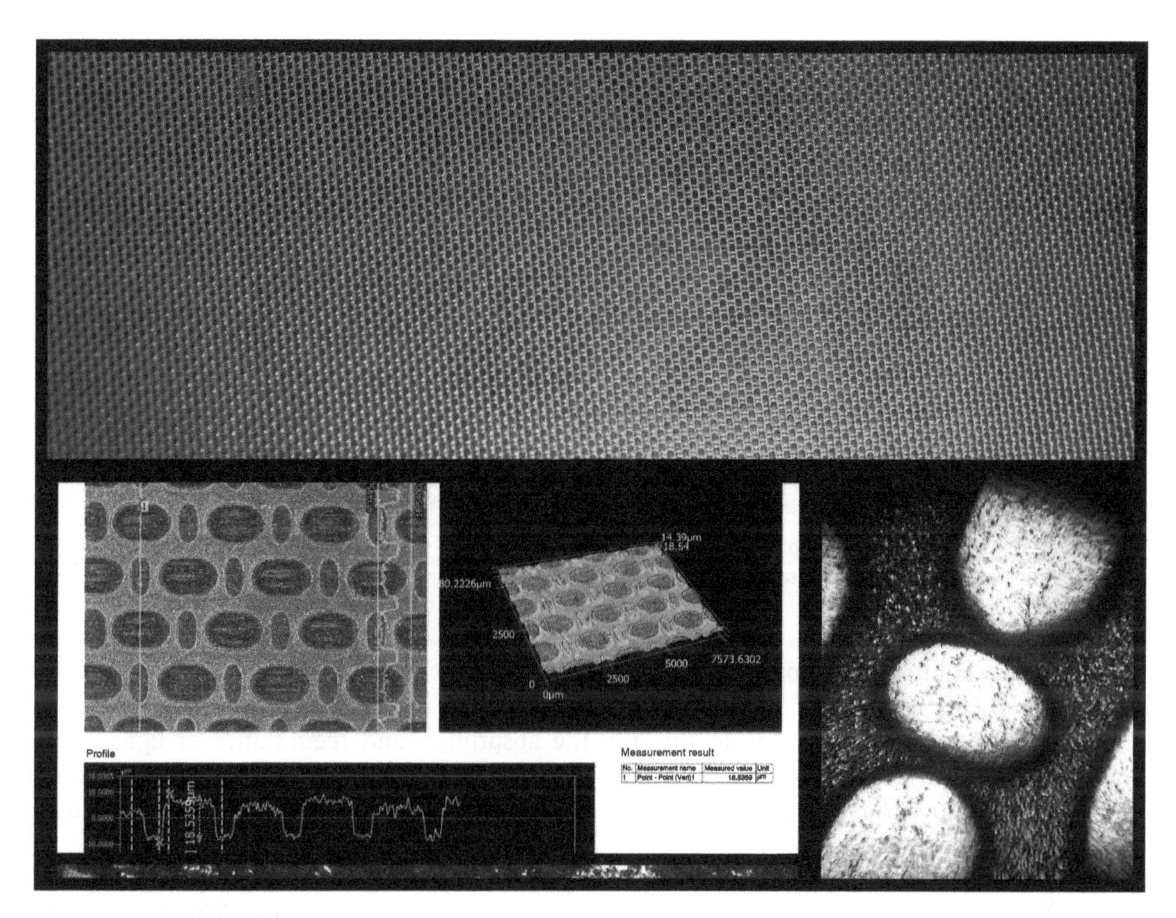

FIGURE 4.19 Linen finish.

Finishes similar to linen that are coined or pressed without bottoming out retain the mill surface on the outward protruding surfaces. When the mill surface is a 2B or 2BA finish, the reflectivity off the tops or the outward bumps is specular. Note the close-up of the linen surface on the lower right-hand corner of Figure 4.19. The tops are smooth mill surfaces, while the lower regions are diffuse. The other regions of the surface obtain their finish from the surface of the high-pressure rolls that impart the finish onto this portion of the surface, making it more diffuse (Figure 4.20).

Unlike paint or plastics, all metals absorb light energy at the surface to a few atoms into the surface. This absorption excites the electrons and causes them to release energy in the form of light. This is the phenomenon that produces metallic luster. With pigmented surfaces such as paints and dyes, an observer can experience the phenomenon known as metamerism referred to previously. Metamerism describes a condition in which the color of two objects appear to be the same under one lighting condition but appear different under a second light condition. Basically, the color we

FIGURE 4.20 Top roll with linen pattern.
Source: Photo courtesy of Rimex Metals.

perceive is the light spectrum of an instance of light coming back to the viewer as it reflects or is emitted from a surface.

Stainless steel is different due to the way the absorption and reemittance of light occurs. Light-scattering surfaces can intensify minor surface irregularities by altering the reflection. Interference will occur as the wavelength of the reflected light interacts with and constructively enhances the reflection.

It is well-known fact that digital cameras are capable of capturing and exhibiting some degree of color difference in metal surfaces, in particular diffuse metal surface reflections. The eye is capable of picking up small differences, but the brain will adjust what we see because our eyes are constantly adjusting to subtle changes. The camera captures an instant, and the sensors and filters many cameras possess tend to enhance small differences in color tones, particularly on perforated surfaces where the light reflected back to the eye or the camera is a combination of the surface reflection and the low reflectivity of the perforated hole.

Color differences will occur. Under different light energies, the perceived color will appear different, and the contrast will often flip as the light energy changes.

The internal variables that will have a determining factor on the color and tone of the metal begin at the Mill source. If the stainless steel used for a particular application is treated using the same heat, annealed at the same time, and finished at the same facility, the finish will be very consistent. If, however, different heats are used, there will be differences in the alloying makeup, and even slight differences will produce a different color tone in the metal that may be visible as the light energy changes. If the stainless steel coils or plates are annealed at different times, the color tone has a high probability of being different. These differences may not be perceivable on close inspection, but as lighting conditions and viewpoint angles change, the differences will be apparent. There is no Mill that will guarantee color tone matches in all lighting conditions. If the metal comes from different cast lots or the heats are different, the chances of variation in color tone are high. Stainless steel is composed of several alloying elements, as described in Chapter 2. The amount of these elements and composition of the surface of the metal as it cools and is rolled into plates and sheets can vary even in a single heat. The atmosphere of the annealing chamber can also affect the color seen on the surface. A cast block of metal is stretched and flattened, and rolled hot and cold until the finish coil is produced.

Different grain sizes that occur when the metal is annealed and cold worked at different times or to different levels will have an effect on the color tone of stainless steel. The different-sized grains will absorb and reflect light differently. These changes are not discernable at close proximity but will be apparent as lighting conditions and viewing distances change.

Mill finishes created as the metal is passed through reducing rolls – or in the case of bright annealed surfaces, polished rolls – can have an effect on the stainless steel's appearance. If the mill rolls are changed before all of the metal is finished, as when the metal is not finished at the same time, the 2B finish can be different and this can carry though as subsequent polishes occur. Mill finish bright annealed surfaces are more refined than others, both in surface quality and grain size. These mill finishes have a better opportunity to match in color and tone when subsequent operations are performed on them.

These variables involve the metal surface characteristics of light scattering and reflectivity, the metal alloy and how it was manufactured in the initial stages of production at the Mill, and the nature and quality of the finish and texture applied to the metal surface. Mills do an amazing job of maintaining consistency in mechanical properties as well as in surface finish characteristics, but there is no guarantee that all portions will match in color tone in all light energy conditions.

ARRIVING AT OBJECTIVE CRITERIA

It is frequently the case that objective criteria need to be established to provide a guide or boundary of acceptability (Table 4.3). Range samples and measurements taken of samples are often the guiding values, but it must be understood that unless these range samples were produced from the actual casting and mill run, they are only representations of what range can occur.

Gloss measurements can be taken by the company producing the finish of the stainless steel surfaces to arrive at a target range. It is standard practice to take gloss readings from different regions of the surface and to arrive at an average reading. This is more common in the paint industry, where the gloss can be established to tighter requirements. For stainless steel finishes, the gloss should be a range. This range may need adjustment once the actual metal is produced. Gloss-reading averages should be obtained along the grain of the metal and across the grain of the metal to arrive at two averages when reading directional finishes such as No. 3 and No. 4. A range can be established by

TABLE 4.3 Objectivity criteria used.

Process	Established control	Limits
Mock-up/prototype using the actual metal for the project	Highest level of accuracy	Little opportunity to change; metal is produced, and high cost has incurred
Range samples from mill	High level of accuracy	Adjustments at secondary finishing operations are possible ways of covering differences
Optical profile	Moderate	Good for physical evaluation but does not often correlate to reflective levels observed
Gloss readings	Moderate	Readings will correlate to reflectivity but not color
Colorimeter readings	Moderate to low	Target colors can vary across a sheet
Luminosity readings	Moderate to low	Changing light source will change readings of energy levels
Roughness Ra	Low	Doesn't always correlate to reflective levels

taking several readings from several sheets or plates. On diffuse surfaces the range across a single element can be significant, yet no apparent difference will be perceivable by the eye. This is because the gloss meters are picking up slight surface disturbances that give rise to the diffuse surface reflection. When using gloss to establish objective criteria for stainless steel surfaces, there should be an acceptable range established and agreed to. A 60° relative gloss meter should be used on diffuse surfaces, while a 20° relative gloss meter should be used on reflective surfaces. Often a Mill will use the 20° gloss meter to read gloss on the more reflective 2B and 2BA mill surfaces, while a finishing company will use a 60° meter to measure the gloss of the applied finish when it is a diffuse surface, even when applied over 2B or 2BA surfaces.

It is not easy to correlate measurements of gloss with differences in appearance. To a degree this works, but there can be other influences that affect the appearance when viewed at a distance or in different lighting conditions. The energy level of the light striking the stainless steel can make some surfaces look dull with respect to the surrounding surface elements, even when the gloss meter readings are well within the acceptable range.

The reflectivity of the stainless steel surface plays a significant function in establishing the character of the surface of the artwork or building facade. How people perceive and interact with the appearance of the surface will be determined by the visual nature of the stainless steel finish.

For colored stainless steels, gloss readings will play a part. Colorimeters also will help in determining and establishing an acceptable color range. These readings are taken at the factory producing the interference or PVD colors on the stainless steel and are part of the quality assurance procedure. On interference-colored stainless steel, the development of the clear oxide that produces the interference is measured with an electrical potentiometer. A given color is associated with the statistical potential resistance generated by the thin film.

FLATNESS

Visually flat surfaces made from stainless steel can be a daunting challenge. There is a distinction between visually flat and measurably flat. To say a surface must be visually flat is an unattainable quality level. The highly subjective nature of a visually flat surface means that "visually flat" is not a specifiable criterion. Attempting to make subjective criteria into a measure of flatness in stainless steel can be like quantifying the blue of the sky. The influence of surroundings, light energy at a particular point in time, and viewing angles will all play a role in how visually flat a surface will appear. Figure 4.21 shows the same grouping of panels. In the top image, the panels are viewed from an angle and appear flat, whereas the lower image is of the same panels viewed directly on. When viewed directly on, the surface shows light and dark regions as the light striking the surface hits at a grazing angle.

Visual flatness is *subjective*. What may look flat in some conditions can look warped and irregular in others. The question is, How do you arrive at *objective* criteria that in the end achieve visual flatness or smoothness in all viewing conditions?

FIGURE 4.21 Distortions in a metal skin are apparent at different angles.

FIGURE 4.22 A flat perforated surface with stiffeners.

What makes attaining visual flatness more challenging for stainless steel than for other materials are two inherent characteristics of the metal. First, there is a significant amount of residual stress in thin stainless steel sheet. The second challenge is the reflectivity of the stainless steel surface (Figure 4.22).

Steps can be taken at the Mill, the finishing facility, and at the fabrication plant to meet established objective criteria for allowable flatness, and yet when the product is installed on the surface of a structure, it still may not appear flat. This is not necessarily because something changed and the piece no longer is flat; it is because the angle of light reflected back to the viewer can enhance the smallest of deviations.

THE TRUE NATURE OF THE METAL

When thin stainless steel sheet is produced at the Mill it undergoes a tremendous amount of pressure as the thin ribbon of metal is pressed between reducing rolls. The center of the sheet sees a different pressure than the outer edges. The ribbon of metal is wrapped around a mandrel in such a way that the center of the coil sees more curvature than the outer edges of the coil. These are real, physical constraints of the production process that cannot be avoided.

The grains that make up stainless steel are elongated and aligned in the direction of the ribbon length. This develops an inherent anisotropy in stainless steel – a condition in which mechanical properties are slightly different in one direction versus the other direction. Thermal expansion will be different along the longitudinal direction versus the transverse, and yield strength will be different as well. This phenomenon is an intrinsic quality of all thin-sheet metal materials (Figure 4.23).

Thin sheets act as diaphragms and exhibit instability across large flat surfaces because the internal stresses within the thin sheet are not consistent. Adding to these unbalanced stresses within thin sheets, fabrication processes can induce uneven stresses into a sheet or remove counterbalancing forces that made the sheet appear flat. Processes of folding or bending can lock the stresses into the body of a panel. For example, folding a pan shape into a stainless steel sheet can both induce stress in the flat, diaphragm portion and lock stress into that portion of the form. The forming of the edge requires the metal along this region to undergo plastic deformation: in other words, the force induced into the sheet to make the bend is sufficient to overcome the internal resistance of the metal and form the edge. Any internal stress at the bend is now replaced by the strength and stiffness imparted by the bend.

Sometimes referred to as "oil canning," wave-like distortions are abhorred in many architectural and artistic applications. This distortion comes from differential internal stresses. Differential

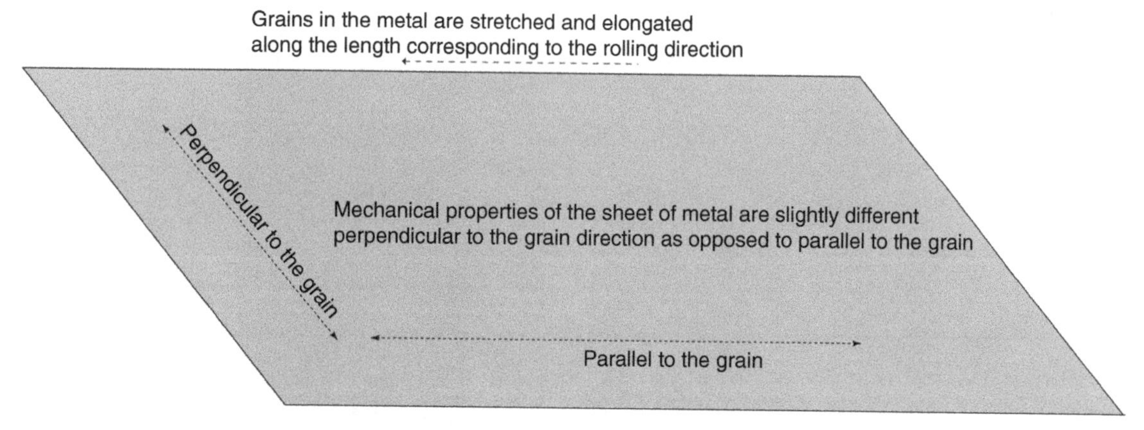

FIGURE 4.23 Anisotropy in sheet stainless steel.

internal stresses can be introduced to the thin ribbon of metal during storage. Coil storage and sheet storage can induce stresses into the metal that will lead to distortions in the fabricated part. The internal distortions can be generated by improper fabrication processes or by temperature changes introduced from welding operations. Generally these internal uneven stresses develop as the metal cools; the cooler parts of the sheet will cool at different rates than warmer regions. However, once the entire surface reaches the same temperature, the wave will often flatten out.

Sheet material produced in coils can have internal stresses that can lead to oil canning in the sheets later on. When coils of stainless steel are produced, they are pulled by force. This can create conditions where the center of the coil is slightly longer. This is referred to as "full center." Full center will be manifested as waves of distortion in the middle region of the sheet because it has been stretched slightly more than the edges of the sheet.

Another, less common coil condition is edge wave. This is where the edges of the coil are longer than the center portion. The edges look wavy when compared to the balance of the sheet. This can lead to distortion in the face surface when the thin metal is formed.

Consider sheet thickness; there is a relationship between the width of a panel and thickness of the material. Additionally, there is a relationship between width and section properties. You might consider these to be the same thing, and in a way they are. If you place a rib in the middle of the sheet or slightly deform the surface beyond its elastic range, you increase the section properties and stiffen the surface. Thicker sheet stainless steel – 1.5 mm to around 4 mm – is usually flatter in appearance. Stretcher leveling of the sheet metal will also help to get rid of some of the internal stresses within the sheet. Stretcher leveling induces even stresses across the sheet by stretching the metal one way and then the other (Figures 4.24 and 4.25).

FIGURE 4.24 Flat 2.0 mm stainless steel in panels 1060 mm wide with ribbed backing.

FIGURE 4.25 Flat 1.0 mm stainless steel in narrow panels 310 mm wide on a private residence.

EFFECTS OF FABRICATION

Perforating or bumping a sheet of stainless steel is an example of how fabrication practices can create internal stress. The process of perforating, or punching, when performed on a CNC controlled press, matches a punch with a die. The metal is firmly held in place as the punch comes down and pierces the metal and pushes the sheared blank of material through the die. This induces internal stress in the area that the hole was created. As more holes are placed in the sheet of stainless steel, the stresses can build up and the sheet will warp. Different techniques can be used to counter the stress to a point, but due to these induced internal stresses shaping will be unpredictable. The thicker sheets, in particular, will warp more when pierced, as the differential stress places uneven tension and compression around the holes. Borders and etches that are not perforated will offer resistance to the localized stretching that occurs in the region that has been pierced. This creates an imbalance that can only be removed by stretching the enter surface. The top image in Figure 4.26 shows the initial warpage in a sheet after the punching process. The middle and bottom images are what the surface looks like after flattening.

Flattening the sheets involves putting them through a process called over-rolling, in which the metal is stretched first one way then another. The warped sheets are passed through a set of rolls to create a large radius and then passed in the opposite direction to flatten them back out. Other techniques involve skipping to other regions of the sheet as the holes are introduced into the metal at the punch press. The stress from one set of holes counters the internal stress generated by another and the surface flattens. The approach is to overcome the internal stress induced from the process of piercing.

FIGURE 4.26 Induced stress in a perforated sheet from punching holes (*top*), then flattened for installation on a bridge (*middle, bottom*).

Press brake forming, where a sheet of metal is plastically deformed in a linear set of tool and dies can induce stress into a flat diaphragm or lock stresses into a sheet.

If the forming operation applies uneven pressure along the length or the bend is not consistent along the edge, additional stress can be generated in the flat, diaphragm portion of the metal. This can occur from inaccurate setting and calibration of the die or from uneven stresses within the sheet (Figure 4.27).

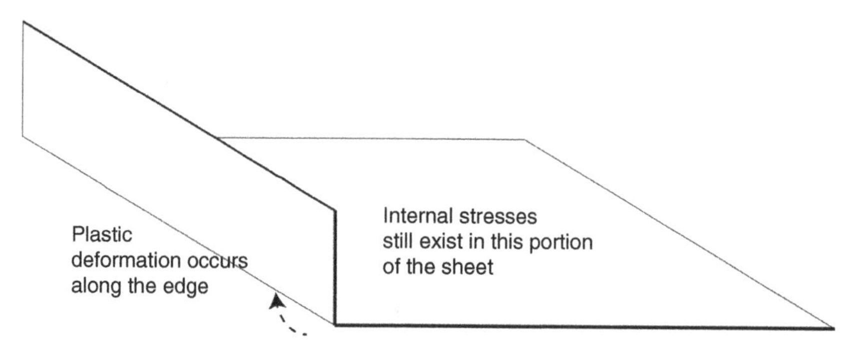

FIGURE 4.27 Stress induced from forming.

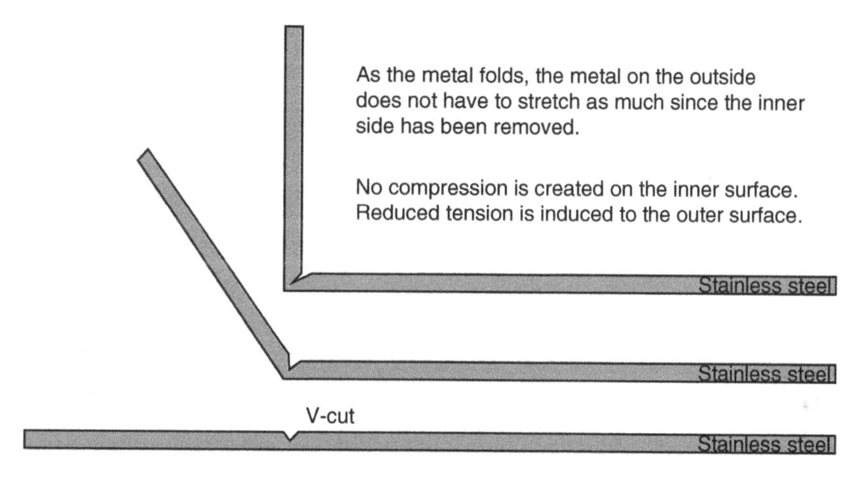

FIGURE 4.28 V-cutting relieves compression on the back side and allows for a sharp bend.

V-cutting the returns has demonstrated a reduction of stress generated from the bending operation to fold an orthogonal return leg. The phenomenon is not fully understood, but it appears to result from relieving the compression at the back surface during folding and allowing the opposite side to see reduced tension forces from the folding operation (Figure 4.28).

The process of v-cutting adds a high level of accuracy to a bend. The bends are sharper and of more precision than a standard bend when performed on thick sheet or plate. V-cutting is a milling process that removes metal without adding a significant amount of heat. V-cut edges can match welded and ground edges, but the benefit is a tighter joint and a reduction of internal stress. The technique of v-cutting stainless steel cannot be utilized on embossed finishes and only for some coined surfaces. The removal of the metal reduces the strength of the stainless steel where the metal has been removed. Stresses should be minimized, and if necessary, the corner can be reinforced with an angle attached with high-bonding tape (Figure 4.29).

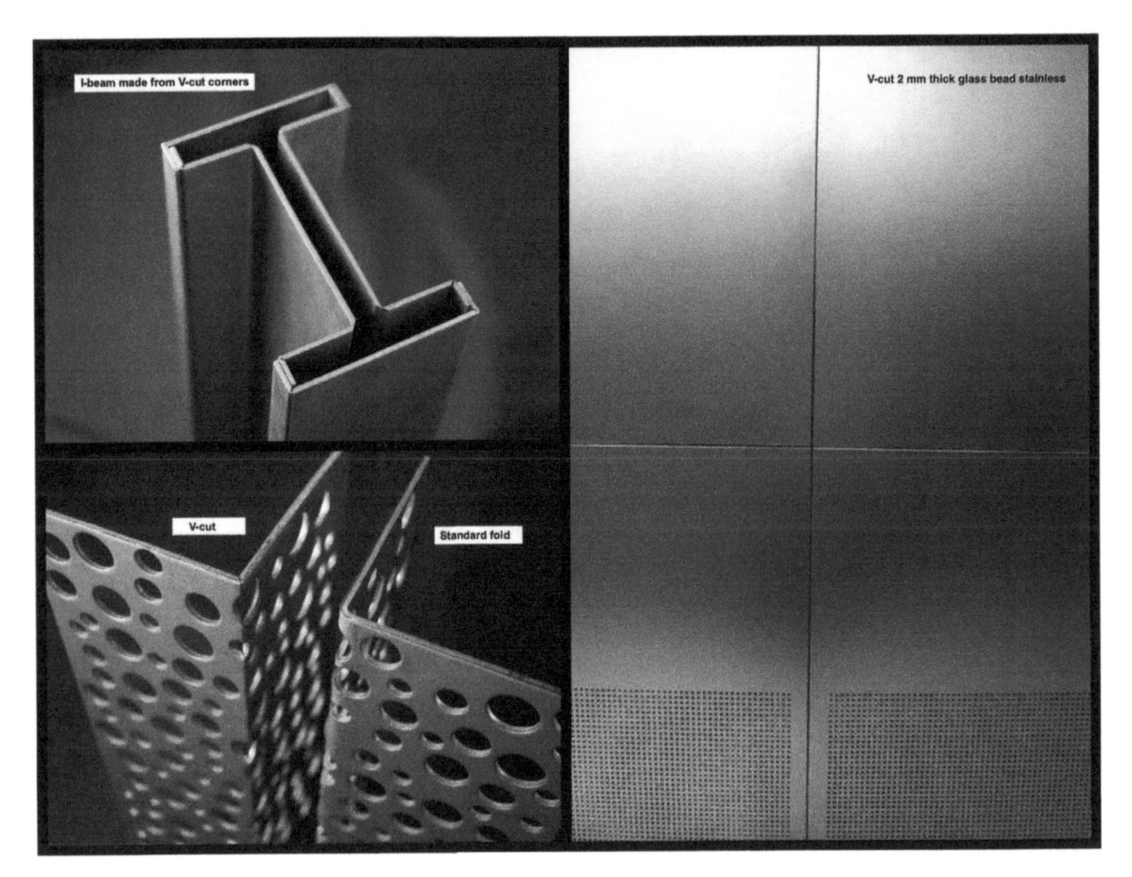

FIGURE 4.29 V-cut edge on 2 mm stainless steel.

EFFECTS OF TEMPERATURE CHANGES

Thin plates act as diaphragms. There is an inherent instability in thin-plate diaphragms. Sheet forms of the metal have an anisotropic character, which implies internal stresses are variable across the flat surface. Temperature changes can add to these internal stresses and create localized warping in and out of the flat plane, as the metal wants to grow outward but is restrained by the fixed edge or the stiffened return.

For thin, flat diaphragms, thermal changes in stainless steels create dimensional changes. If the edges are confined by forming a return, thermal changes can cause movement in the central region of the sheet such that portions may be concave while other portions are convex.

Often on thin metal roofs made of stainless steel, distortion (descriptively called "oil canning," as previously mentioned) will be more apparent in the late afternoon sun as the surface heats up and the metal wants to expand. Grazing light angles will enhance minor differences. When the plane is a flat diaphragm, the metal expansion will be both convex and concave, even on

TABLE 4.4 Thermal coefficient of selected stainless steel alloys.

Alloy	Coefficient of thermal expansion (10^{-6} m/m °C)
S41000	9.9
S30400	17.3
S31600	16.0
S32205	13.7

the same surface. Because of the reflective nature of stainless steel, this undulating surface is more apparent. If, however, the diaphragm is restricted to moving only outward when the surface expands, the distortions are less apparent because they no longer manifest as an undulating wave but as an outward pillowing.

To determine how much movement is to be expected from temperature changes, the coefficient of thermal expansion is used. The coefficient changes depending on the stainless steel alloy used. Table 4.4 shows that the austenitic alloys (S30400 and S31600) will expand and contract at a greater rate than the martensitic (S41000) and duplex (S32205) alloys.

To determine how much thermal expansion to expect or to design for, use the following formula with the table of coefficient of thermal expansions for similar alloy types.

$$\Delta L = L_i \times \partial\,(t_f - t_i)$$

ΔL = Change in length expected
L_i = Initial length of part
∂ = Coefficient of thermal expansion
t_f = Maximum design temperature
t_i = Initial design temperature

For example, assume that the stainless steel panels are 3000 mm in length and 1020 mm in width. The panels are delivered to the project site in the winter months and the ambient temperature is 4°C when they are installed. Also, summer temperatures can be expected to reach as high as 35°C.

The stainless steel is a S30400 and the coefficient of thermal expansion is 0.0000173.

Plugging this information into the formula, we find that the panels can be expected to expand as follows:

$$\text{Length of panel: } \Delta L = 3000 \times (0.0000173)(35 - 4) = 1.60\ \text{mm}$$

$$\text{Width of panel: } \Delta L = 1020 \times (0.0000173)(35 - 4) = 0.55\ \text{mm}$$

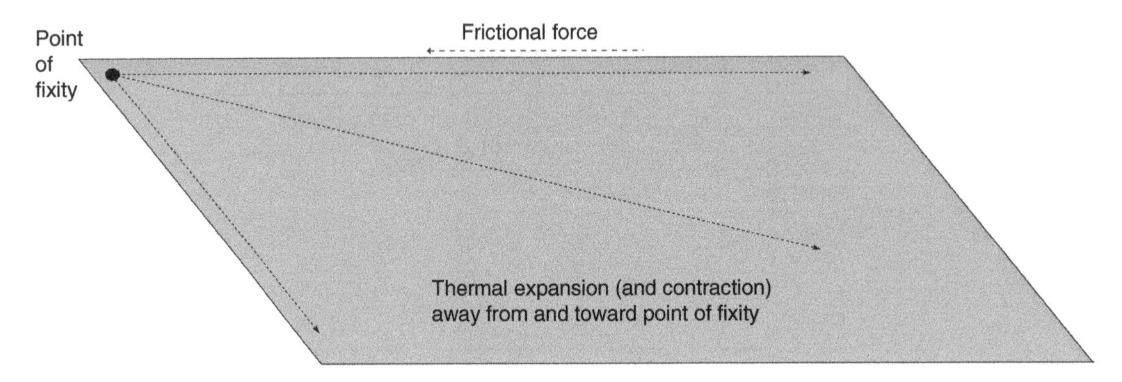

FIGURE 4.30 Thermal expansion away or toward a point of fixity.

Therefore, the design needs to allow for a minimum of 1.6 mm of thermal expansion to occur in the lengths of the panels and 0.55 mm in the width. Where this has an effect on the stainless steel panel is its ability to "push" the thin plate away from a given fixed point. If the section of the metal is not sufficient, it will create a surface distortion. Some refer to this as a "buckle," but this is not the case. A buckle is a permanent plastic deformation in a surface. Here the surface distortion is not permanent, and as the temperature changes the occurrence and appearance will change (Figure 4.30).

Stainless steel is stronger than many other metals used in art and architecture. Formed edges are stiffer and stronger. But if the panel is not allowed to move as thermal conditions change, distortions will appear in the center of the panel where the diaphragm is less stable.

Welding stainless steel will induce heat into the surface and can cause significant localized distortions around the welds. Welding is discussed in depth in Chapter 6. Once distortions are introduced from the heat of welding they are permanent, because the solidifying metal fixes the stresses into the metal. To weld stainless steel, temperatures of 1450–1500°C are necessary. These temperatures will expand the surrounding stainless steel at different rates as you move away from the metal.

Stainless steel requires an understanding of the metal and welding skill. A stainless steel piece can be permanently damaged if proper procedures are not followed.

When welding stainless steel, heat is more concentrated: unlike aluminum or copper alloys, stainless steel does not conduct heat away as efficiently. The heat-affected zone (HAZ) around a weld is where most of the distortion will occur. As the metal is heated, it expands, and when the weld solidifies, this can fix the now-expanded metal so that a wave or distortion is now locked into the piece. However, there are ways to reduce distortion when welding stainless steel. The use of chill bars, inert gas, and skip welding techniques can reduce stress built up from the welding process (Figure 4.31).

A welded surface or edge can be indistinguishable from the body of the piece. Figure 4.31 shows a welded assembly with zero distortion. To produce this, an aluminum heat sink was used while the weld was flushed with argon. The joint was welded using gas tungsten arc welding (GTAW) attached to a robot.

FIGURE 4.31 Welded edges without distortion at 200 11th Avenue, New York, designed by Selldorf Architects.
Source: Photos courtesy of Selldorf Architects.

If your work has a significant amount of welding, consider using one of the low carbon alloys so as to reduce the development of chromium carbides that form under the tremendous heat of welding. The use of chill bars in order to reduce the heat by pulling it away from the HAZ is critical for good results when welding stainless steels.

THE RELATIONSHIP BETWEEN THICKNESS AND FLATNESS

There is a correlation between width of a panel, surface finish, and visual flatness. Visual flatness simply means that the surface appears flat. In the art world this might be sufficient, but in architecture objective criteria are often called for in the specifications used. The problem is that they are often meaningless because they lack basis. The specification may call for a measure of flatness with no real correlation to how the surface will appear. The challenge is to find objective criteria that can be achieved consistently in order to provide an acceptable visual appearance in

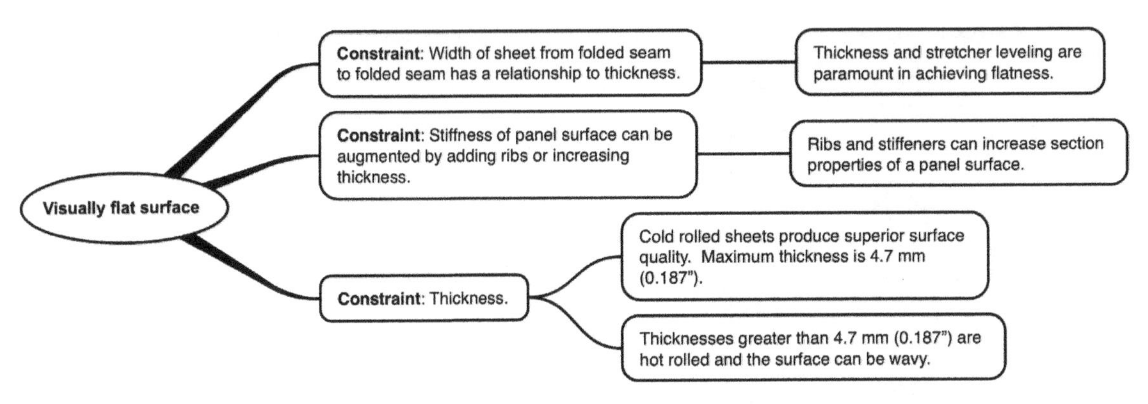

FIGURE 4.32 The relationship between width and thickness.

all viewing conditions. In the end, it is how the surface appears to the viewer that matters. Using measurements based on arbitrary values without taking into account surface finish, angle of view, viewing distance, and temperature will only lead to frustration. What measurement is sufficient to produce a visually flat surface? This cannot be an arbitrary condition and should be worked out with the supplier and fabricator (Figure 4.32).

Figure 4.33 shows the relationship between the seam to seam of a panel made from cold rolled, annealed stainless steel. It identifies a range of nominal thickness measurements and relates those

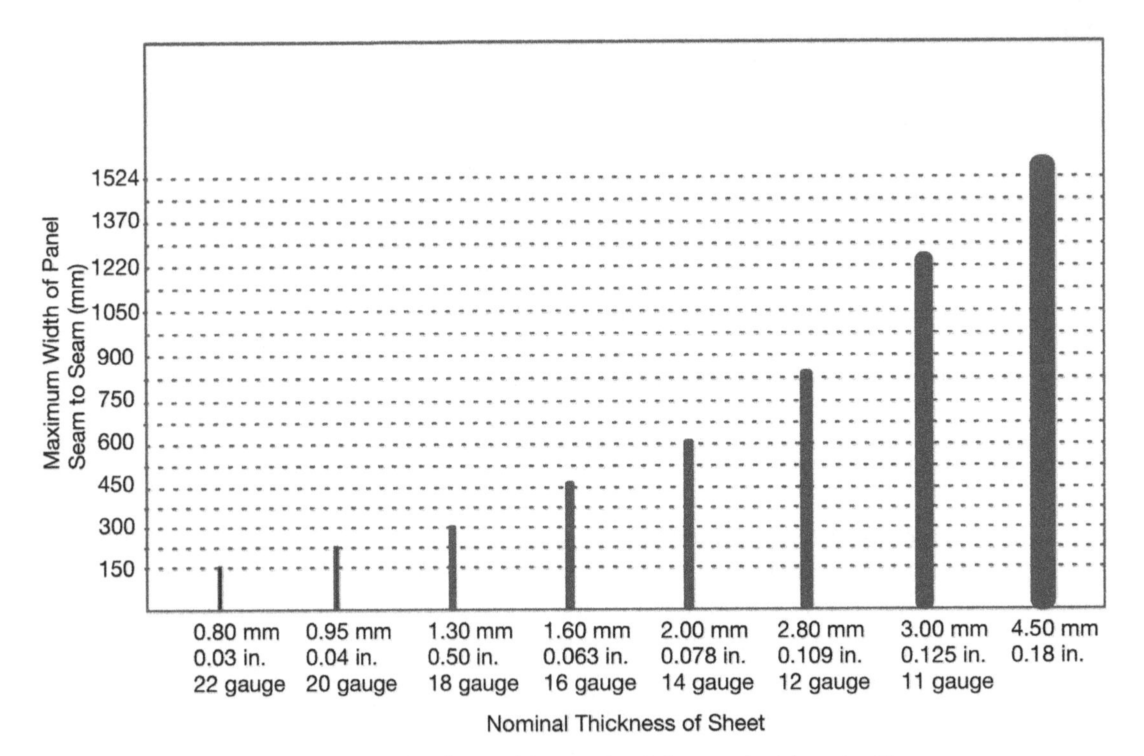

FIGURE 4.33 Maximum panel width for sheet thickness of stainless steel without backing.

measurements to the maximum seam-to-seam spacing that can be used to achieve a visually flat surface in an unbacked material. Seam to seam is any formed edge where a panel is made by folding inward or outward of the plane of the metal surface. Please note that the ratios shown assume good stretcher leveled metal and medium reflectivity. For highly reflective surfaces, the width of panel should be reduced by 30%.

Essentially, what the chart shows is there is a strong relationship between the stiffness of material and the stability afforded by the diaphragm created by the unsupported metal surface. As the width increases, there is a need to increase the thickness of the metal and therefore the stiffness and strength of the metal if you wish to achieve flatness.

The ratios shown in Figure 4.33 can also be used to space stiffeners on the reverse side; however, the process of placing the stiffeners can induce deflections in the face side of the sheet. The process is not the same as applying a bend or rib in the body of the surface, and it can be difficult to arrive at a sheet or panel that is precise in flatness and leveling. The pressure used to place the stiffener must be even and consistent. The adhesive, fusion stud, or other means used must be applied consistently and accurately: any change, and you will lock in distortion. The process of attempting to place a stiffener into a panel can create distortions in the final product as the adhesive cures. Once this happens it is very difficult to correct.

PROTOTYPE VIEWING PROTOCOL

When creating flatness by forming or by addition of stiffeners or backing plates, produce a prototype. Review the prototype using an angle, distance, and lighting condition that have been previously agreed upon. Sometimes the surface will look appropriate from one viewing angle under a certain lighting condition and look horrible in another. Recognizing the effect of grazing lighting conditions and the way reflective surfaces can enhance minor changes in plane are the first two steps in gaining an understanding of how the surface will appear. Matte stainless steel surfaces may appear dull in relation to other stainless steel finishes, but they still are more reflective than most other materials (Figure 4.34).

Often the requirements of a project state that the surfaces must be examined when viewed from a distance of 3 m. The surface distortions and the tonal differences may not show until viewed from a greater distance, and often at an angle.

TEXTURING TO IMPROVE FLATNESS

Texturing the surface by embossing or rolling patterns into the sheet will assist in achieving a visual flatness. Some of these processes create localized stretching on the surface that permanently deforms the metal into the pattern dies. They are very effective methods of overcoming the internal stress within a sheet of material. They will also contribute to making the surface more diffuse. Embossing is performed on sheet thicknesses that are generally in coil form, but it can be induced in sheets before forming them into panels or shapes (Figure 4.35).

Another method is to induce a slight outward camber in the face of a flat panel. There is a bit of an art to this, because too much creates a pillowing appearance and too little opens the surface

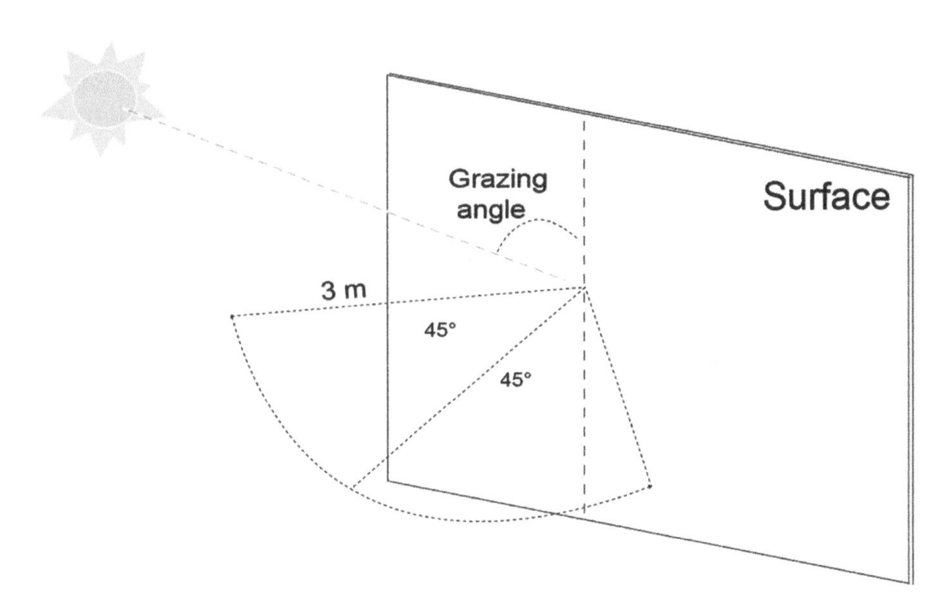

FIGURE 4.34 Establish a viewing angle and criteria up front.

FIGURE 4.35 Embossed and interference colored stainless steel on a Nieman Marcus store, designed by Elkus Manfredi Architects.

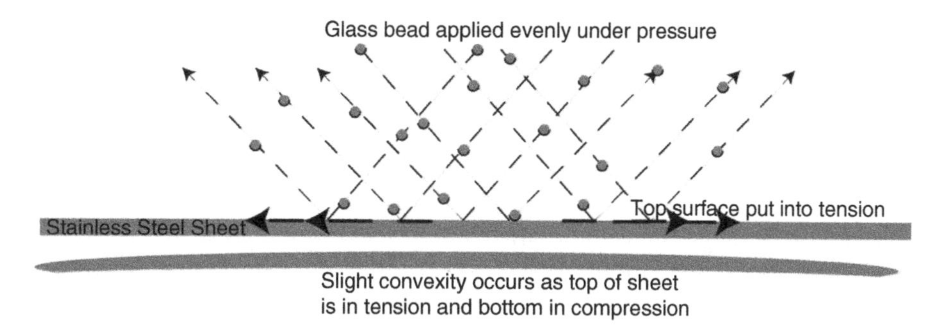

FIGURE 4.36 Shaping the surface of stainless steel sheet.

into showing more distortion as thermal changes cause portions of the slight curve to go the other way. Light glass bead blasting one side (the side you want to bend outward) can induce a surface with a slight bow. Where the bead blasting strikes the surface it undergoes stretching with respect to the nonblasted surface, and thus you have a stress imbalance that bows the sheet. It is important to use automated or semiautomated blasting equipment. Using handheld devices will impart uneven and overlapping blasting patterns that will only intensify the unwanted surface distortions (Figure 4.36).

STANDARDS OF FLATNESS

In 1955, the American Iron and Steel Institute (AISI) commissioned Princeton University to carry out a study that would determine a measure for flatness. The study looked at stainless steel panels in various finishes used at the time of the study. Some of the panels were backed with stiffeners and others were not. Princeton arrived at seven contributing factors that influence the visual appearance of stainless steel panels:

1. Minor production variations at the Mill source.
2. Sheets that are not flat when shipped from the Mill.
3. Unequal stresses resulting from fabrication. Local stresses may be imposed by forming and welding; formed edges of sheets may be work hardened while the center is still in an annealed state.
4. Stresses resulting from the erection of panels. Only a slight amount of racking, warping, or squeezing of a panel when it is installed could be sufficient to cause waviness.
5. Expansion due to temperature rise.
6. Shrinkage of the backing can have the same effect as expansion of the face.
7. Movement of the building frame can rack or warp panels that were originally set plumb and level.

The study went on to say that some panels initially installed and found to be acceptably flat later developed waves. The objectionable distortion was apparent in both unstiffened panels and those with stiffeners. The study assumed that this was due to thermal changes, but since when the temperatures returned to previous levels the waves remained, it concluded that the thermal expansion only accentuated the waviness that already existed. It also concluded that the waviness could influence the internal stresses that were in the sheet.

Flatness of a panel is a geometric tolerance where measurements are taken of a surface as it compares to a true theoretical counterpart. Sometimes the tolerance is given as some predefined value. For the Mill source, that value can be quite significant if left to industry standards. For the most part, Mills that supply stainless steel fully understand that undue distortion will create havoc in the polishing, fabrication, and design world, so they work to produce a high-quality surface to very tight tolerances. Sheets should be properly leveled to achieve an even distribution of internal stresses across the surface. Current standards for Mill sources of stainless steel define limits for the height of a wave from a theoretical flat surface. It does not define the number or frequency of the wave. You could potentially have multiple ripples that fall within the standard but create an aesthetic issue. Figure 4.37 shows a stainless steel panel 3 mm thick with stiffeners along the length and a custom perforated surface. In the right-hand image the panel appears visually flat, but when measured with

FIGURE 4.37 Measuring flatness.

a straight edge, the panels show a significant wave. Stiffeners in the panels run the long length, which does nothing to flatten the cross-wave. The v-cut edge and fold flatten the perimeter of the panel, but the center has an inward bowing wave due to the perforation pattern, which stretches the metal slightly.

Specification terms such as "panel flatness," "special flatness," and "half-standard flatness" are meaningless without a definition, as the criteria used to define flatness is arbitrary.

At the Mill source, the specifications covering flatness tolerance are varied and generous. Such standards, if applied to finished art and architectural surfacing elements, would never meet visual flatness from a subjective perspective.

For example, ASTM A568/A568M-17a is the specification covering the general requirements for steel sheet in coils and cut lengths. Table 20 ("Flatness Tolerances of Cold-Rolled Sheet Cut Length") in the specification states that for thicknesses over 0.044 in., on sheets 36–60 in. wide, the maximum deviation from a horizontal flat surface is 0.375 in. for 45 KSI yield steels. Additionally, there is no limit to how many such deviations can occur in a given sheet. This is the allowable flatness for steel sheet from the Mill supplier. This is a significant bow. Most designs would find this unacceptable. Most fabrication facilities would struggle with this amount of deflection.

The standard goes on to reference another method of describing flatness. It is called the I-Unit. The I-Unit method takes into consideration the height and frequency of occurrence of each wave along the length of the rolling direction of the sheet material.

The lower the I-Unit, the flatter the material will be to start with.

The I-Unit is determined by laying the sheet on a flat table and measuring the frequency and height of the wave.

The formula is:

$$\text{I-Unit} = 2.467\,S^2 \times 10^5$$

where

$$S = \frac{\text{Amplitude (height of wave)}}{\text{Wavelength (spacing of wave)}}$$

Thus, using the formula, one would measure the occurrence of the wave in a given sheet. The constant, 2.467, is derived from $(\pi/2)^2$. Both the height off of a flat horizontal and how often the wave occurs in the sheet of material are measured. This is better than what ASTM presents in its table. Still, a maximum value for the I-Unit needs to be established and agreed upon. Keep in mind that a flat sheet resting on a flat surface will have gravity working in its favor. The weight of the sheet will work to reduce the measurement of the amplitude (Figure 4.38).

On formed panels, sheets with edges are turned to produce a 90° turn along the edges, meaning the waves will be locked in. As previously mentioned, this condition is described as oil canning.

So, how to arrive at the I-Unit acceptable for art and architecture (Table 4.5)? According to this calculation, a 3 mm rise in a panel of 1524 mm (or slightly less than 1/8 in. in a panel of 120 in.) would give an I-Unit of 0.96. Per the algorithm, this would give a similar flatness value to a 1.5 mm rise every 760 mm along a panel. *In surfaces used in art and architecture, the I-Unit should be less*

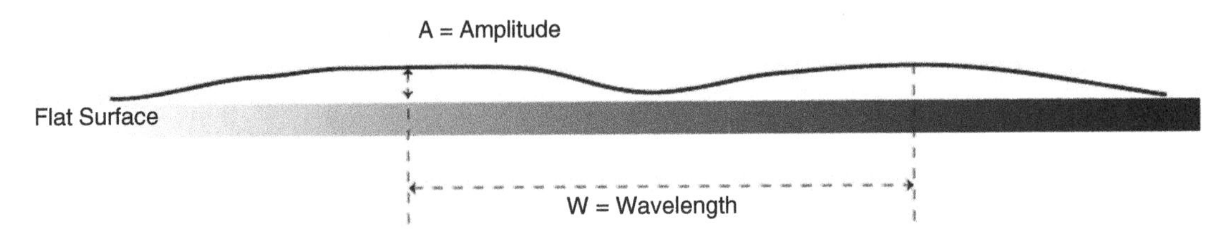

FIGURE 4.38 Measuring the amplitude and wavelength.

TABLE 4.5 Selected I-Unit values for different amplitudes and wavelengths.

Amplitude (A) (mm)	Wavelength (W) (mm)	I-Unit
1.5	760	0.96
1.5	1020	0.53
1.5	1524	0.24
2.0	760	1.71
2.0	1020	0.95
2.0	1524	0.42
3.0	760	3.84
3.0	1020	2.16
3.0	1524	0.96

than this value to achieve visual flatness in most architectural finishes and in all viewing angles. This is very difficult to achieve, and particularly challenging if the sheet you begin with does not conform to this guideline.

Another method of measuring the angle created by the distortion in the surface relies on putting the surface in the plane it will exist in. This method considers the slope of the surface but misses the amplitude the I-Unit value captures.

Figure 4.39 shows two ways of measuring flatness in a surface. This method has been used as a guide for decades. Case I covers situations in which a panel has multiple waves, and Case II covers those in which the panel has a single bow. Originally, back in the 1960s, the percentage allowed was much more generous: as much as 1.5% for dull surfaces and 1% for reflective surfaces. Keeping within the values of 0.5% and 0.25% shown in the figure will make for a flatter surface, but regrettably, the surface will still not always be visually flat.

Depending on the finish, lighting conditions, and angle of view, Case I will still appear with contrasting light and dark areas on the surface of a single panel. Whether it meets the threshold of visual flatness or not is dependent on the subjective position of the viewer. Measurable criteria may

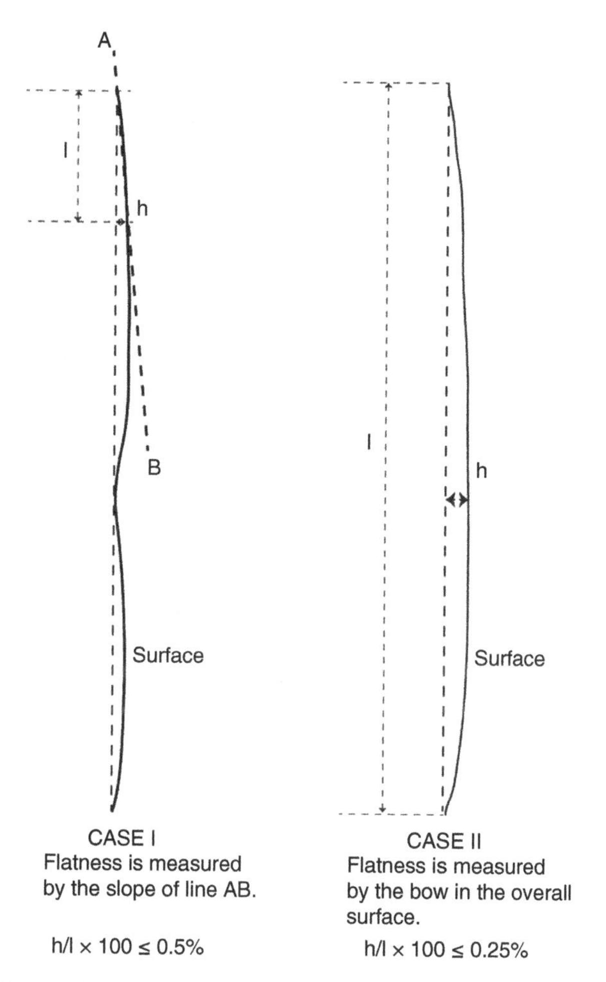

FIGURE 4.39 Two conditions and corresponding measures of flatness.

meet the specification requirements. Close-up inspection in the plant or on a prototype mock-up may meet visual approval. But when installed and viewed from multiple angles and under variable lighting conditions, the waves can be apparent. Once in place, there is nothing that can be done to bring it into 100% visual acceptance.

Case I is what often occurs when stiffeners are applied to the back surface of a panel. Thermal movements can accentuate the rhythm of the distortions echoing from the stiffener placement. In most cases, it is recommended to push panels to Case II. Reflectivity plays a major role. The more reflective the finish, the more the slightest distortions will be apparent. However, even with light-scattering finishes, such as glass bead and Angel Hair, the contrast of light and dark levels can manifest in certain viewing conditions (Figure 4.40).

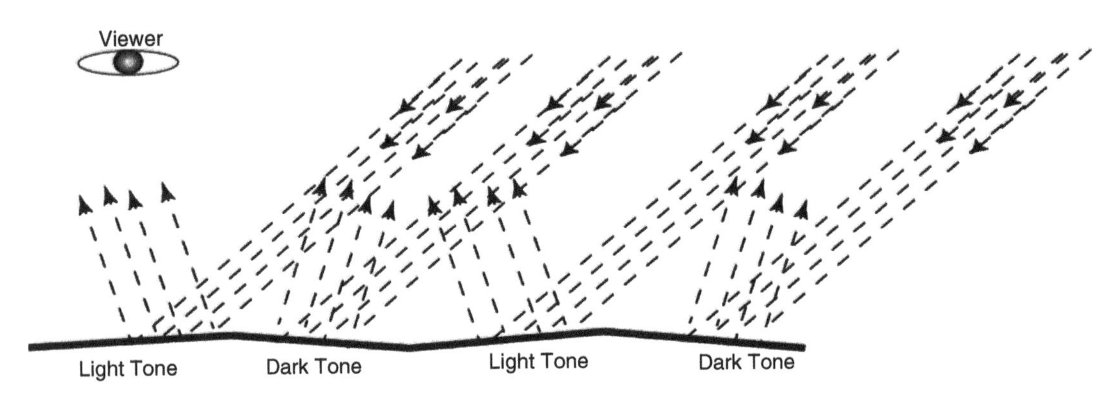

FIGURE 4.40 Tonal differences from reflection in a Case I situation.

In a Case II situation, the panel has a consistent bow. The bow is outward, not inward. Convexity is a requirement. Concave surfaces will concentrate the reflected light inward and the contrasting light and dark zones will be exaggerated. Convex surfaces reflect light outward. On diffuse surfaces that scatter light, the metal on either side of the joints or intersections of the panels may appear in contrasting lighter or darker regions – but only slightly. The eye reads the consistency of the light reflection off the consistent bow and does not pick up visual distortions.

For meeting visual flatness or smoothness thresholds, one needs to go to extremes in analysis of how a surface will appear. Prototypes that follow each step in the process – finishing, manufacturing, handling, and installation – should be planned, developed, and viewed to identify all issues that will affect success. Simply stating it must look flat or that it must look smooth does nothing and is a non sequitur. In all industrial criteria flatness is objective, and the surface may very well meet accepted industry standards but fail visual acceptability.

GRAIN DIRECTION

Grain direction will play a part in the ultimate appearance. The direction of the grains is established in the cold rolling process. In sheet material the grain direction runs the length of the sheet. This is because when the sheet is manufactured as a large ribbon of metal at the Mill, the grains are worked and stretched along the length of the ribbon. Expansion and contraction are greater in the direction of the grain than across the grain. There is a slight bias in reflectivity as well, due to the direction of the grain. When a panel is formed, the bends across the grain cold work the stainless steel at the bend point. The grains are stretched slightly more than the grains in the middle of the sheet. This can induce internal stresses into the flat area and add some instability to the diaphragm. Folding the edges stiffens the area around the fold, but it can also trap internal stresses in the center of the now-formed sheet. The center of the formed element will often exhibit local warping in and out of plane. Depending on the finish, this warping may not be visually apparent up close. A straight edge set over the surface will show slight in and out of plane shaping. Grazing light over the surface may show the enhanced distortions as light and dark zones, which will often become apparent from a distance of several meters.

To achieve a flat surface, consider the following:

1. Involve the Mill source and establish an acceptable I-Unit, which should be no greater than 0.50.
2. Insure the packaging is in good crates (solid base, not slats).
3. Involve a quality fabricator. This is key. The equipment must be calibrated for working with stainless steel. The fabricator must be versed in how to weld stainless steel to prevent weld distortion.
4. Create prototypes of full-size panels for viewing in various lights and angles.
5. Establish a finish that is diffuse and light scattering. Relative gloss range should be established as a criterion.
6. Use a thickness appropriate for the width of the panel. Do not sacrifice thickness, otherwise you will sacrifice quality.
7. V-cut edges if possible. V-cutting removes the compression stress at the bend and does not contribute to stress or distortion inboard of the bend.
8. If width exceeds thickness limits, consider:
   (a) Fully backing the surface
   (b) Building in slight concavity if possible and eliminating Case I conditions
   (c) Adding a shallow die-pressed design if possible, which will break up the reflection and stiffen the surface
   (d) Setting stiffeners very carefully if they are used, insuring that the laminated supporting area is large, and considering a slight concavity if possible
9. Utilize a measuring system to validate that no waves or ripples exceed 0.25%.
10. Design a support system that will allow the panel to move and expand without racking or distorting as building movement occurs and as thermal movement in the metal occurs.
11. Involve an installation company that understands how to handle and install stainless steel panels without inducing stress.

Many designers and fabrication facilities place stiffeners across the surface to stiffen the diaphragm, but all this does is break the distortion into a series of smaller distortions, and it can cause a series of contrasting waves as well. Placing the stiffeners along the length of the diaphragm can work and it is often less visible, but it can still lock the differential, in and out of plane, to areas between the stiffeners. Figure 4.26 shows perforated panels with stiffeners along the length rather than across the panels. Applying a stiffened corrugated sheet or a honeycomb backing are two ways to increase the section and provide support to the diaphragm. Stiffeners are efforts to make up for thickness reduction. If not very carefully applied, they can create a condition worse than if no stiffeners were ever installed.

The thicker the sheet, the better. There is no substitute for thickness if keeping a surface flat is the requirement, even when stiffeners are incorporated on the reverse side of the surface. The use of stiffeners and the thickness of the stainless steel sheet go in tandem. If the thickness is inadequate, the stiffener only compounds the differential stress in a flat diaphragm by creating localized constrained regions. If the thickness is sufficient and the stiffeners are applied consistently, the surface can approach flatness in most lights (Figure 4.41).

FIGURE 4.41 The Wayne L. Morse United States Courthouse in Oregon, designed by Morphosis. The Angel Hair finish is 2 mm thick.

If stiffeners are needed to induce flatness in a surface, it is recommended that a slight outward camber be introduced. As the surface expands or contracts, the change in geometric form is on the induced radius of the face. If you set the stiffeners as close to flat as possible, there remains an instability as the surface moves. More often than not, the thermal changes in the surface will be both inward of the plane and outward of the plane. This undulation will be more visible in glancing light across the surface. Inducing a very slight curvature can be done by adding a slight curvature to the stiffener, or better still inducing a curvature in the sheet surface and in the stiffener. The edge fold will work against this curvature, because at the edges the diaphragm comes back to flat. There is no good way around this, so it is critical that the curvature is very small; otherwise, you begin to initiate a pillowing appearance.

Adhesives that involve a thermal transformation for activation can create visible waves on the face side when viewed in certain lighting conditions. It is recommended to use double-sided tapes. They provide a more consistent adhesive surface for a metal interface than tube-dispensed adhesives. They can be used together to fix the stiffener into place accurately and evenly while the tube-applied adhesive cures. Unevenly applied liquid forms of adhesive can create inconsistencies in the appearance on the opposite side.

GRAIN SIZE AND SURFACE CLARITY

There is a point in the manufacturing of stainless steel sheet and plate where the production is hot rolled and does not undergo cold rolling operations. This happens for production of plates greater in thickness than 5 mm, or 0.187 in. The hot rolled stainless steel plate's surface is rougher and less refined. The grains are not as elongated and are larger than cold rolled stainless steel. The thinner the sheet material, the finer the grain size, and a finer grain size is needed to achieve a good surface appearance. Fine grain sizes receive a polish much better than large grain sizes, and the only way this is achieved is by cold working and annealing the sheet.

Finer grain sizes:

- Improve surface appearance
- Improve surface polish
- Improve strength
- Improve corrosion resistance

You can expect less flatness and occasional flaws, such as surface inclusions, in hot rolled material. These inclusions cannot be polished out and appear as darkened pits in the surface. Often, the more you polish, the more they appear. Finishes are more difficult to apply to hot rolled stainless steel because of its flatness, grain size, and surface inclusions: mirror surfaces are cloudy and satin

FIGURE 4.42 An example of a 6-mm stainless steel plate with a light glass bead blast surface.

finishes are not consistent. Electropolishing the surface will brighten clarity, but electropolishing works much better when the grain size is finer.

With plate, you can achieve a finish that is superior and has a quality that is more refined yet tough. The Mill source will need to run a cold pass on the plate. This gives the surface a smoother appearance. A light Angel Hair or glass bead blast can induce a matte quality into the surface. On close examination the surface will have a refined grain appearance with subtle inconsistencies (Figure 4.42).

There are several ways to influence the grain size in stainless steel. The alloys that contain titanium and niobium have finer grains. These added elements restrict the growth of the grain. Low carbon forms of stainless steel have the opposite effect. The reduction of carbon leads to larger grain growth.

The smaller the grain size, the better the clarity and the better the polish and brightness of the metal. Electropolishing, for example, works best to brighten stainless steel surfaces when the grain size is at least an 8, or a maximum of 0.022 mm in size (Table 4.6).

ALLOYING CONSTITUENTS AND SURFACE CLARITY

Sulfur content in the alloying elements will have an effect on the finish clarity of stainless steel. Sulfur is added to improve machining. Some high-speed welding operations perform better with a higher sulfur content. Pipe, bar, and welded tube alloys will usually have a higher sulfur content than sheet or plate material, but sometimes the Mill may produce coil stock from castings

TABLE 4.6 Grain size.

ASTM grain size no.	Size (mm)
1	0.25
2	0.18
3	0.13
4	0.09
5	0.065
6	0.045
7	0.030
8	0.022
9	0.016
10	0.011
11	0.008

designated for pipe. Sulfur makes it difficult to polish the surface because it can create miniscule pits and small indentations. The more you polish, the more these small pits become exposed to the surface. For good clarity and finish, have the Mill provide stainless steel with a sulfur content of less than 0.005%. You can remove sulfides from the surface using a passivation technique outlined in ASTM A967, but this adds cost and complexity. If possible, work with the Mill source.

Chromium content on the high end of the range can improve the surface clarity of stainless steel. Molybdenum and nickel improve corrosion resistance, but they can have a slight dulling effect on the finish of satin or matte surfaces.

There are simple methods of determining the precise alloying constituents on stainless steel surfaces. One common technique is used by the scrap industry in determining the probable alloying content of the metal. This phenomenon is X-ray fluorescence, or XRF. XRF is a nondestructive method of analyzing the surface of stainless steel and other metals to determine the alloying makeup. XRF measures the fluorescent X-rays emitted from a sample surface when it is bombarded by another X-ray source with the XRF gun. Each of these emitted X-rays is a unique signature of the element on the surface. The gun reads the X-rays and compares the results to a stored database. The results provide the statistical probability of one alloy versus the next, based on industry criteria of the alloy makeup. They are very accurate qualitative analysis tools and will accurately provide criteria for the makeup of a stainless steel surface (Figure 4.43).

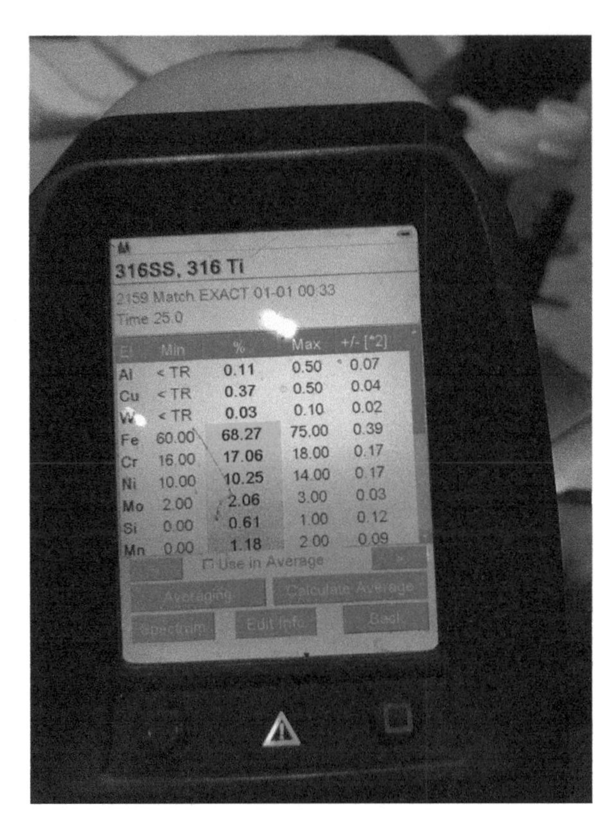

FIGURE 4.43 An XRF reading of a stainless steel surface.

EXTERNAL VARIABLES: ENERGY OF LIGHT SOURCE

The energy contained in different lighting conditions varies depending on the source. Energy reaching the earth from the sun will vary depending on the time of day, the season, and the atmospheric conditions. Daylight spectral distributions, a measure of the power of the wavelength hitting an object, change throughout the day along with intensity. In the early morning and late afternoon as the apparent elevation of the sun decreases, the sunlight will pass through more ozone due to the increased length of its path. This will filter out the greens and yellows. This period is called the "blue period" in photography. In contrast, the greatest level of radiant intensity develops in the period between late morning and early afternoon.

Stainless steel and its component alloys absorb this light and reemit it based on the particular atomic makeup of each component. Tiny variations, coupled with the surface finish and its microscopic profile, can make stainless steel appear less glossy or more yellow in certain light energies. At night, when illuminated by artificial light that has an energy profile different from sunlight, the surface may look patchy. Panels that might have looked lighter in the sunlight will now look darker than surrounding panels.

This is the nature of natural metal surfaces such as stainless steel. At times all natural metals will exhibit this tendency: copper, aluminum, titanium, and zinc can all have a similar patchwork character. The surfaces are not paint. There is not a pigment that absorbs a given wavelength and reflects others on a consistent, predetermined basis. Stainless steel surfaces react and change with the light striking them. For some, this is the beauty of natural metal. Metallic paints attempt to copy the luster and beauty of a natural metal only to look plastic and unnatural.

The key to a large stainless steel surface is not to place contrasting surfaces in large groups within a field of other, contrasting surfaces. Achieving this can be extremely difficult, because the small discrepancies in the metal and the metal surface reflection are not apparent or discernable at close inspection. The only means is to manage the coil, bar, or plate material from the Mill through the fabrication facility and on to the building. Even when this is done, some differences in tone can still manifest in certain lighting conditions (Figure 4.44).

Inspecting for color tonal differences can pose a significant challenge for several reasons. At the fabrication facility, sheets and coil material often arrive with a protective coating on the surface. This is desirable to prevent scratches and mars on the surface, but it also hides the appearance. Peeling the film and inspecting the surface of each can be significant in terms of cost and time. Even then, if the metal is not viewed in certain light only major differences will be uncovered. The fabricator of the form or panel will often work with the protected sheet and add significant labor and fabrication cost. When the color is revealed at the project, major cost has gone into the surface and changes can be expensive. To overcome these challenges, the fabricator and Mill source must have a good relationship and understanding of quality assurance processes to arrive at the highest value for the end product. This may drive the cost of the Mill product up by making more stringent quality

FIGURE 4.44 Interference colored stainless steel after peeling off the protective plastic and discovering several panels of a different color.

practices the Mill's responsibility, but these costs are far more economical than trying to adjust after fabrication.

Early in the morning and late in the afternoon the angle of the sun can be at a grazing angle and exaggerate the level of distortion in the stainless steel surface. This will also occur on some surfaces where the sun angle is high and the view is from underneath the surface. It is common to inspect flat surfaces by illuminating the surface at an acute angle, sometimes referred to as a grazing angle. This exaggerates flaws in the surface and shows as contrasting shadows. Very small deviations in surface flatness will be exposed and accentuated when the light passing across the surface does so at a very acute angle.

EXTERNAL VARIABLES: ANGLE OF ILLUMINATION

Building surfaces composed of multiple reflective elements can make for an overwhelming sight. Each element may show only minute deviations from flatness, but when installed on a wall, angles of view and distances of view can accentuate the most minute differences. One panel may appear flat while the adjoining panels show waves. As the viewer moves this changes as the angle of reflection from the surface changes. If the surface is a diffuse surface, expect a more enhanced affect as the light waves bouncing off the surface interact and generate or alter the wavelength returning to the eye of the viewer. Differences in inflection points on a stainless steel surface can be as small as a millimeter or less, yet the distortion can appear as several times as large when observed in grazing light. This effect is particularly enhanced when the surface is concave rather than convex. On a concave surface the light reflecting back to the viewer is concentrated to a focal point. As the viewer moves this point changes.

SURFACE FINISH AND PERCEIVED DISTORTION

The reflective nature of stainless steel will magnify minor differences in localized flat metal surfaces. There are two basic types of reflection from a stainless steel surface: specular and diffuse. Both types of reflection play a role in achieving a flat surface. Specular reflections are one to one, mirror-like surfaces that can show distortions immediately as the observer perceives changes in the reflected image. Highly reflective surfaces show changes in light intensity as well, where one section is bright and glowing while another is dark. Specular surfaces are unforgiving and show the slightest distortion. Even curved surfaces made from reflective stainless steel are challenging to achieve optical clarity. *Cloud Gate*, the sculpture by Anish Kapoor, is the best example of painstaking care taken to minimize distortion in a stainless steel surface (Figure 4.45).

Directional satin finishes, such as No. 4, can demonstrate a reflective level at which achieving visual flatness can prove difficult. They are more reflective in one direction than another and can show differential surface curvature across a single sheet. Up close they may appear flat, but as the observer moves back, surface undulations become apparent.

FIGURE 4.45 *Cloud Gate*, designed by Anish Kapoor.

This phenomenon is a challenge for diffuse reflective stainless steel surfaces. Visually they appear very flat when observed up close. Even within 3–4 m they can appear visually flat. As you move away surface appearances begin to change, partially due to the light-scattering behavior of the surface. Herein lies the relationship with the blue of the sky. The sky appears blue in color due to light scattering: particles in the sky scatter the sunlight, and because blue light waves are shorter, they are scattered more intensely. When the sun is overhead, the blue of the sky is apparent from this light-scattering behavior. When the sun is low, the blue light is scattered further and the blue color is removed.

With diffuse textures on stainless steel, there is a distance effect that can play into how smooth or irregular the surface will appear. Surfaces may look flat when observed at close proximity, but when viewing the surface from a distance, reflective differences will appear. This phenomenon is based on changes in focal point. Up close, the reflected scattered light waves are more intense and localized. The observer up close to the surface perceives a flat surface because of the intensity of the scattering. When the observer is farther away, the intensity is reduced and the surface abnormalities

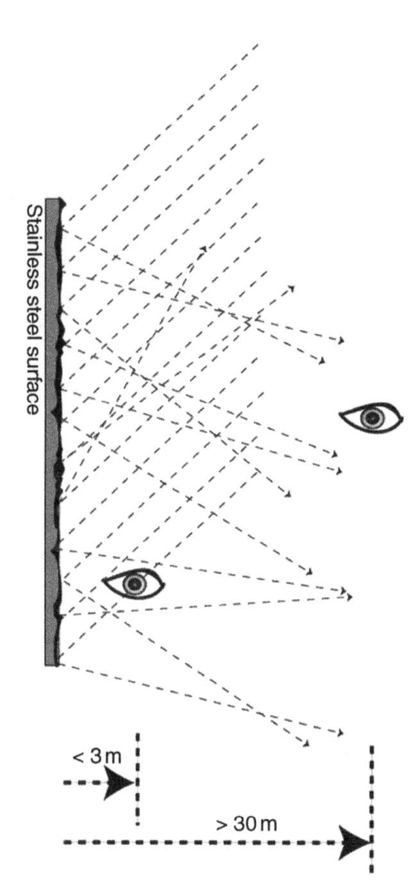

FIGURE 4.46 Distortions become apparent at different viewpoints.

and variations in light intensities can become apparent, making the surface appear to be more distorted (Figure 4.46).

In diffuse surfaces, grazing light (light striking the surface from an acute angle) will make small irregularities in the surface more apparent. Grazing light creates shadows and intensifies irregularity in surfaces. Surfaces can look smooth and flat in most light conditions, but once the light moves to strike the surface at an acute angle, irregularities intensify.

ARRIVING AT THE BEST POSSIBLE OUTCOME

Taking into account all the variables that will affect the appearance of wrought stainless steel used in art and architecture, the following steps can help insure success:

1. Select the best quality metal for the environment.
   - Use austenitic stainless steels, or if the chloride environment is severe, one of the duplex stainless steels.

- Use low carbon alloys and require sulfur content to be less than 0.005%. Controlling the sulfur to low levels will aid in arriving at clarity in the final finish. The low carbon content will allow welding processes to be easier to clean and refinish.
- Use metal from the same heat. Avoid, if possible, mixing heats. If this is not possible, then plan out the use of different heats on different elevations and different surfaces.
- Use metal that has been annealed at the same time when possible. If this is not possible, then plan out the use of the different batches on different surfaces as much as is practical.
- Use stretcher leveled sheet or plate material and establish a method of inspection with the Mill supplier to remove all coil stops and chatter marks in the sheet or coil before the material is covered with protective plastic.
- Protect the stainless steel with the proper paper interleave or plastic coating. Use laser coating material if the metal is destined for laser cutting. Mark a direction arrow on the protective coating to indicate the direction in which the stainless steel was produced into coil form.
- Ship the metal in coils with protection wrap and edge protection. If the metal is decoiled into sheet or plate form, the crates must be free of any nails or projections into the metal and must support the entire length and width of the stainless steel.

2. Handle and prepare the finish carefully.
   - Apply the surface finish in the same grain direction as indicated on the sheets or coils.
   - Inspect the stainless steel for defects and flatness criteria. Establish a quality method to insure the flatness of the material before fabrication steps.
   - Establish a method to insure appearance criteria is achieved. Monitor color, gloss, and reflectivity. Arrive at a range of acceptable gloss.
3. Produce a mock-up of the surface of sufficient size to demonstrate to the designer and client the most probable appearance.
   - The mock-up or prototype should be made from material of close proximity to the material and finish intended for the final surface.
   - Arrive at an acceptable viewing angle and distance.
   - Leave the mock-up in place for reference until an acceptable surface is applied to the structure.
4. Establish measurable objective criteria to use as a guide, not an absolute.
   - Until the actual material is produced, range samples established from floor material may not represent production stainless steel in color and tone.

The goal is to arrive at an acceptable surface that meets design and aesthetic criteria by careful preparation and planning.

CHAPTER 5

Designing with Available Forms

Engage people with what they expect; it is what they are able to discern and confirms their projections. It settles them into predictable patterns of response, occupying their minds while you wait for the extraordinary moment – that which they cannot anticipate.

Sun Tzu, *The Art of War*

Stainless steel has barely a century of industrial usage, yet it is considered one of the most important design materials known. Stainless steel, an alloy of steel, possesses properties that provide both mechanical and aesthetic characteristics unmatched by any other material known to mankind.

Attributes of Stainless Steel

- Metallic luster
- Corrosion resistant
- High yield strength
- Durable
- Malleable
- Accepts custom finishing
- Does not tarnish in the atmosphere
- Accepts interference colors
- Accepts physical vapor disposition colors
- Easy to clean
- Specific alloys can be tailored for specific uses

- Extremely long useful lifetime
- Castable
- Can be recycled
- Accepts electropolishing

BASIC FORMS OF STAINLESS STEEL

Stainless steel comes in various forms that are similar to the forms available for steels. The process for creating these forms begins in the same way but then branches off into production sequences specific to that form. The forms all have various limitations due to the initial constraints of production and subsequent constraints imposed by manufacturing processes.

As discussed in Chapter 2, there are two base forms of stainless steel: wrought forms and cast forms. You could throw in a third form and call it beads and powders, but these are typically either used to create a finish or melted down into another form; in addition, they are derived from one or the other of the base forms.

Wrought Forms

- Foil
- Sheet
- Plate
- Bar
- Pipe
- Tube
- Wire
- Extrusion
- Structural

Cast

- Sand castings
- Permanent mold casting
- Investment casting (lost wax)
- Ceramic mold

Shot, Powder

- Cast stainless steel shot
- Cut wire stainless shot
- Fine powder

THE STAINLESS STEEL MILL

All forms of stainless steel begin at the Mill source where the alloy is created. The Mill producer takes charge of recycled scrap and adds in various alloying components – for example, nickel, chromium, and nitrogen – as well as other alloying elements. These are melted to industrial specifications and alloy tolerances in an electric furnace. Once melted, the molten stainless steel is cast into one or several basic forms: blooms, billets, slabs, and rounds. The melting and initial cast is called a heat. A heat will generate a mill certification with a specific heat number for that specific casting of metal. A tracking number is assigned and imprinted on the initial forms. Metallurgic analysis is undertaken on a sample of the heat to verify alloying constituents and physical properties and the corresponding specification – American Society for Testing and Materials (ASTM), American Society of Mechanical Engineers (ASME), Deutsches Institut fur Normung (DIN), British Standards (BS), or other – to assure the standard specification has been met. Figure 5.3 shows a mill certification form with the information usually provided. Often accompanying the form is a description of the heat treatment used as well as other information associated with various requirements and specifications.

Mill certificates monitor and document the stainless steel produced in a given heat. They follow the stainless steel form as it goes out to stocking houses and the fabricator or surface finishing company of the metal. The mill certification is an important document in the context of quality assurance. The heat may be broken down into different forms, slabs, or billets and then further rolled into sheet, plate, wire, or rod, but the certification information follows the material and is accessible to the user for verification. The modern Mills of the twenty-first century melt the stainless steel in an electric arc furnace and then move the melt to the AOD chamber where argon-oxygen decarburization improves the characteristics of stainless steel by removing carbon and sulfur. From here the molten stainless steel is continuously cast into slabs, billets, or blooms (Figure 5.1).

The slab form is reheated and passed through hot rolls to thin and elongate the slab into a thick ribbon of black hot rolled stainless steel. The surface of the thick hot rolled and pickled stainless steel plate is ground, and the edges are trimmed. This thick ribbon is coiled into large rough coils and placed into an annealing chamber to soften the metal. The coil is further hot rolled and passed through a pickling line. The clean thick hot rolled ribbon of metal is recoiled or flattened and cut to length as hot rolled plate. The hot rolled and pickled coil often is sent to be turned into sheet or strip. In the case of the stainless steel shown in Figure 5.1, the rolls of thick stainless steel will be sent to a Sendzimir rolling mill, which will pass the ribbon under a set of small diameter rolls with a tremendous amount of pressure applied. The metal will be rolled cold through the Sendzimir rolling mill. The Sendzimir rolling mill, also referred to as a Z-mill, thins the ribbon to the thickness specified. As this happens the grains elongate, and internal stresses develop. The metal will then be passed through an annealing oven. Modern systems use a continuous annealing line.

At this point the now-thin ribbon is very bright and clean of oxides. The metal is at a particular temper and possesses defined mechanical properties. The metal can receive a skin pass to produce 2B or 2D mill finishes or it can be sent through a bright annealing line to produce the 2BA finish. The ribbon is either recoiled and stored as coil, slit to smaller strip coils, or decoiled and sheared to arrive as sheets skidded to lengths.

FIGURE 5.1 A hot slab of steel at a South Korean stainless steel Mill.
Source: Photo courtesy of POSCO.

From the Mill the metal is sent to a stocking warehouse or to an end user. In either case it is provided to specification and the mill certification will travel with the order.

A fabricator and manufacturer of stainless steel parts and assemblies for art and architecture should accept only the best surfaces from the Mill. Stainless steel should have scratch- and pit-free surfaces with the specific mill finish applied to the surface. The surfaces should be free of scale and oxides and have a consistent appearance. They should be well protected, dry, and clean.

Mills will produce large plates to specification for various industries. The plates, sometimes referred to as mill plates, are large: several meters in length and as much as 2.5 m in width. The Mill will also produce coils of common and popular thicknesses used in industry. These are kept along with tubes and bars at intermediate processing warehouses as inventory for sale. The alloys are determined by market needs and the finishes are normally the mill finishes (Figure 5.2).

For material normally destined for the architectural, transportation, and food-service markets, quality is maintained and the surfaces are protected. For heavy plates and bars used for general industrial applications, the finish is not as critical and the surface quality and alloying constituents may not be suitable for architectural and artistic use. For example, some heavy plate has a higher than normal sulfur content to better facilitate tube and pipe fabrication in high-speed welding operations, which benefit from the higher sulfur content.

When ordering stainless steel from a stocking warehouse, it is important to know what the nature of the surface will be and the alloying constituents. The higher-sulfur alloy for high-speed welding, for example, is still considered to be the S30403 alloy, only with sulfur on the high end of the allowable range.

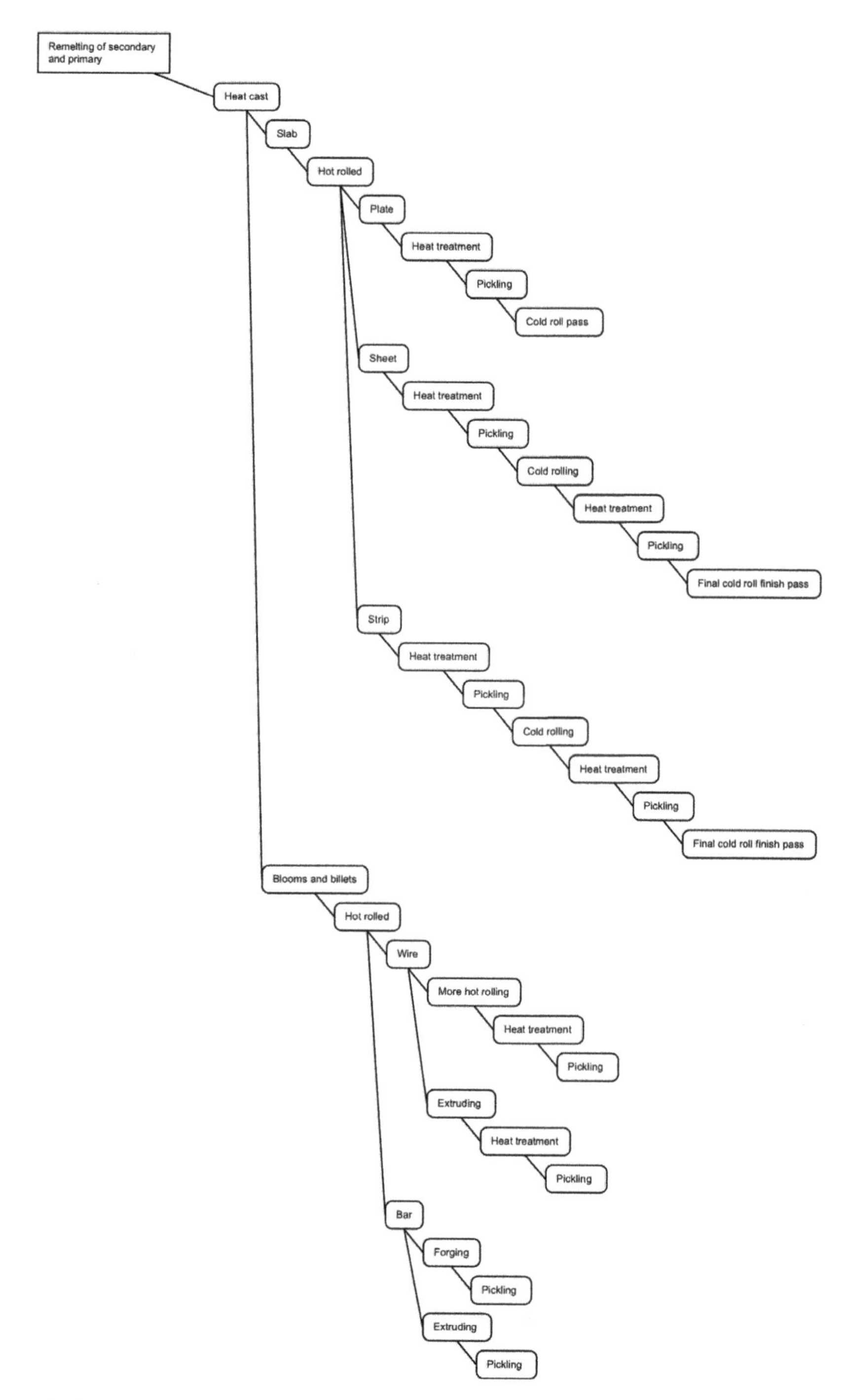

FIGURE 5.2 Basic forms produced at the Mill source.

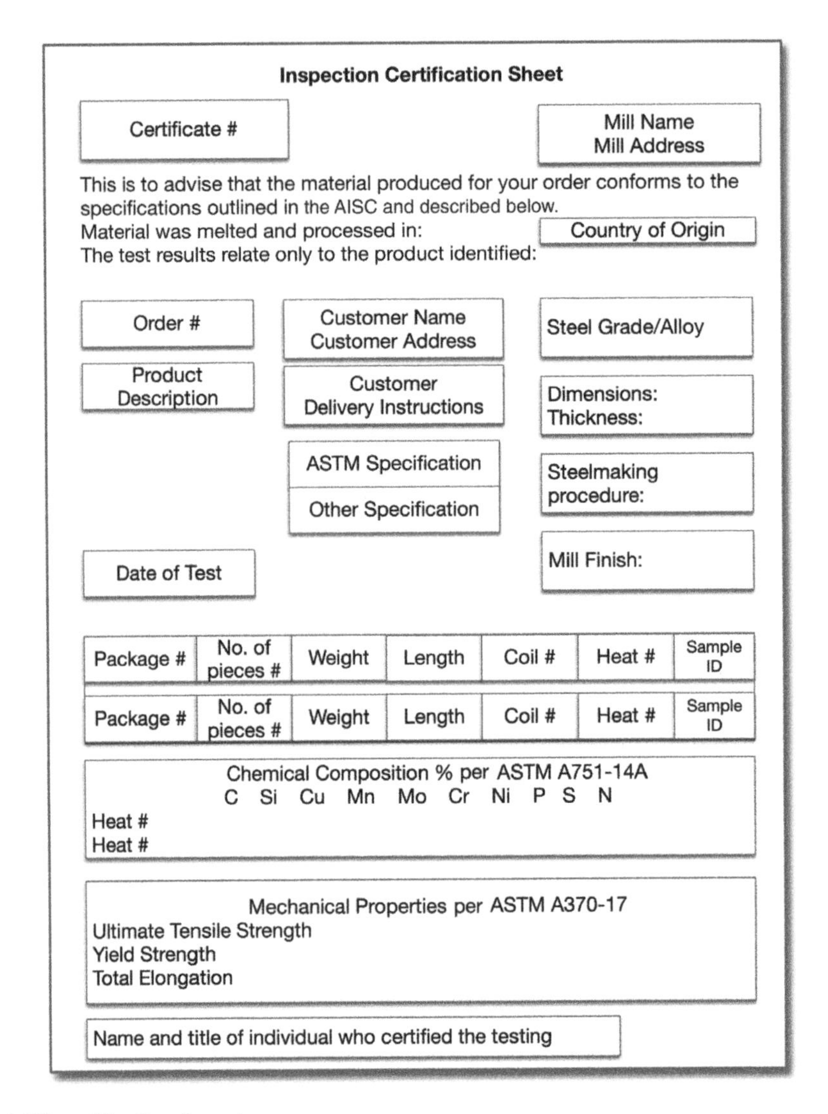

Inspection Certification Sheet

Certificate #

Mill Name
Mill Address

This is to advise that the material produced for your order conforms to the specifications outlined in the AISC and described below.
Material was melted and processed in: Country of Origin
The test results relate only to the product identified:

Order #

Customer Name
Customer Address

Steel Grade/Alloy

Product Description

Customer Delivery Instructions

Dimensions:
Thickness:

ASTM Specification

Other Specification

Steelmaking procedure:

Date of Test

Mill Finish:

Package #	No. of pieces #	Weight	Length	Coil #	Heat #	Sample ID
Package #	No. of pieces #	Weight	Length	Coil #	Heat #	Sample ID

Chemical Composition % per ASTM A751-14A
C Si Cu Mn Mo Cr Ni P S N
Heat #
Heat #

Mechanical Properties per ASTM A370-17
Ultimate Tensile Strength
Yield Strength
Total Elongation

Name and title of individual who certified the testing

FIGURE 5.3 Mill certification format.

Figure 5.3 shows the information that a mill certification form contains. It has three main sections. The top identifies the Mill that produced the original casting of the metal and includes the original certificate number that will travel with all forms of the metal as it moves from the first solidification form, such as the billet or slab. The next section identifies the ordering entity and the governing specifications of the order as well as a description of the form and the finish the metal was ordered to. This section could indicate, for example, that the metal form is coil, and it would specify the weights of the coils and the heat number that the product was created in. The heat number is important; it is the stamp of a particular melt at a particular time.

The last section of the mill certification form indicates the chemical makeup of the material and its mechanical properties. The mechanical properties can relate to a temper used to achieve the particular strength properties. Sometimes for coils the mechanical properties of the lead-in portion of the coil and the end portion of the coil will be listed. Essentially the Mill shears off a section at the beginning and the end of the coil and tests their strength and measures their elongation. There should also be a reference and a stamp or a signature of which facility and which metallurgist certified the results.

This is established practice and an important assurance about the metal ordered for the project. Without this certification, you cannot be certain of what you really have unless you perform an in-depth analysis of the material. If something does not appear right, you can trace the problem back to the source.

MILL PACKAGING REQUIREMENTS

All fabricated stainless steel mill products, plate, sheet, coils, rods, and wire are carefully handled. It is a different process from that at a conventional steel mill because of the nature and expense of the metal. Stainless steel has a quality expectation higher than that for steel, and as such must be handled accordingly or the characteristics that enable the metal to be of special distinction will be damaged or altered.

The skids used to support decoiled plate and sheet are solid, flat, and nonmetallic. Fasteners used to hold the skids together cannot and should not be protruding. Usually a layer of protective Masonite or stiff cardboard is set on the top of the skid below the first sheet or plate. Corner protectors are used, and other protective coverings are employed at points of potential contact. For coils of stainless steel, the edges are protected and the coils are encased in a protective paper wrap. Rods and bars are stacked and protected to eliminate scratches from handling. Nonmetallic bands are used and at no point is steel in contact with the stainless steel.

Pipe and some hot rolled plates may be rough and given an "as fabricated" surface finish. These are usually destined for an industrial process that does not concern itself with the aesthetic, only the strength, durability, and corrosion resistance. However, they should be protected from steel banding and have the edges and sides protected from scratching, marring, or gouging (Figure 5.4).

Sheet and coil destined for architectural and art fabrication should have a protective layer on the finish surface. This is usually in the form of a very thin plastic film. Many films today are laser ready–type films to allow for laser cutting operations. A direction arrow that indicates how the metal was taken off the coil is usually included on the protective film. This is good practice and the direction indicator should follow from the beginning fabrication process to the end installation.

Transportation and Storage

- Dry
- Surface and edge protection
- No steel bands in contact with the stainless steel

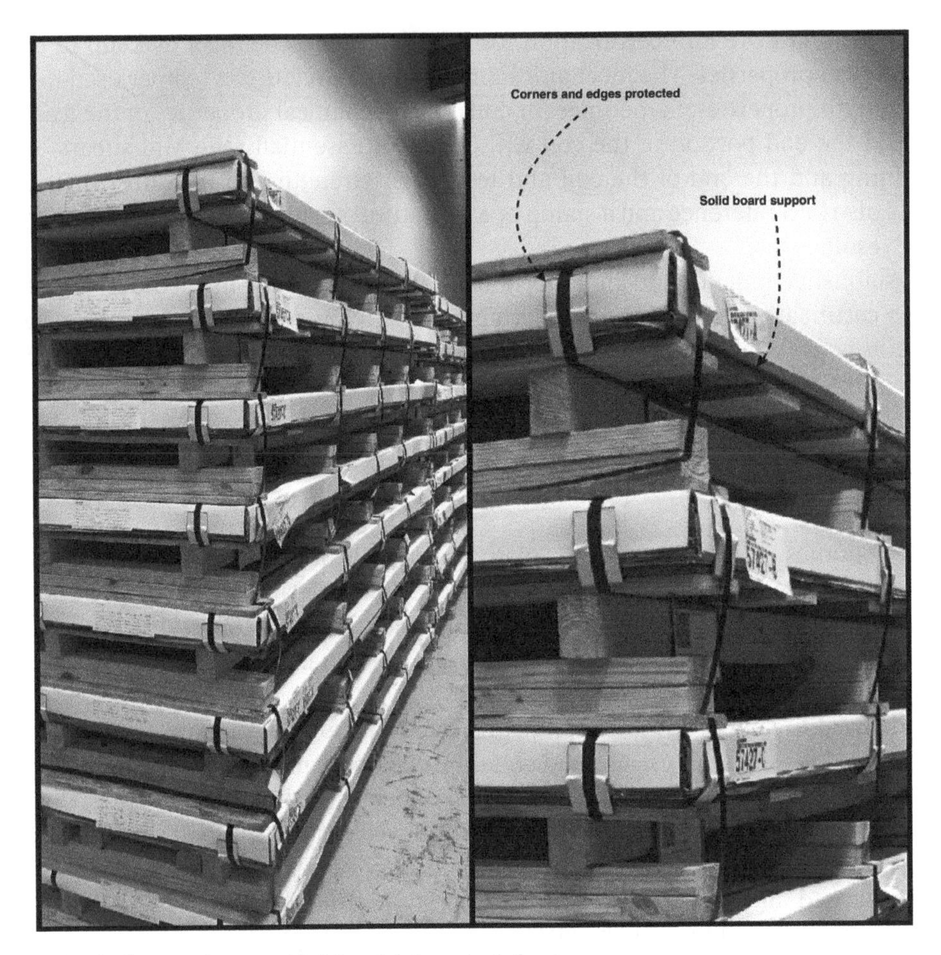

FIGURE 5.4 Typical protection provided for stainless steel sheets.

- Flat, smooth, strong pallets
- Coil edges protected
- Full wrap of coil and sheet pallet stack

SHEET: THE ARCHITECTURAL FORM OF STAINLESS STEEL

The most commonly used form of stainless steel in art and architecture is the sheet form. Large sections of the stainless steel Mill are dedicated to the production of stainless steel sheet. For most stainless steel Mills, it is the most important, defining form of the metal they produce. Because of this it receives the most quality assurance attention. Sheet production and the final finishing of the

TABLE 5.1 Stainless steel sheet thicknesses.

Millimeters	Inches	Closest gauge
5.00	0.197	7
4.00	0.157	8
3.00	0.118	10
2.50	0.098	11
2.00	0.079	12
1.50	0.059	14
1.20	0.047	18
1.00	0.040	19
0.80	0.032	20
0.60	0.024	22
0.50	0.020	24
0.40	0.016	26
0.30	0.012	28

sheet form are undertaken in a relatively clean portion of the plant. Sheet material is usually coated in a protective film as it is decoiled, leveled, and cut to length. The equipment used to produce sheet forms of stainless steel are monitored and adjusted on a regular basis to insure proper working order. Stainless steel sheet production is one of the most stringent quality assurance undertakings of industrial material processing. Stainless steel sheet production is on the border of art and science.

Stainless steel sheet is considered thicknesses of 4.76 mm (0.1875 in.) or less (Table 5.1). It is important to designate a minimum, maximum, and nominal thickness in inches or in millimeters. Gauges are not defined by industry standards and the descriptive term "gauge" for stainless steel is now relegated to industry jargon. Those in the industry may refer to a 14-gauge thickness, but this is understood to be a nominal 1.5 mm thickness. Gauge definitions are not a good way of specifying stainless steels. Designate the thickness in inches or metric equivalent. Most Mill producers are moving to the metric designations.

STAINLESS STEEL SHEET DIMENSIONS

Nominal Thicknesses

It is important to note there are tolerances in these values, both plus and minus from the nominal thickness. Mill producers want to maximize the production of material, so they push to the low end of the allowable range that the industry has established. That being said, the thickness of stainless steel should not fall below the minimum range established by industry. ASTM A480 establishes US standards for allowable thickness tolerances. These are aligned with widths of sheet material. The wider the sheet, the greater the tolerance allowed. For example, on a stainless steel sheet 2 mm thick, widths up to 1.3 m will have a plus or minus thickness tolerance of 0.1 mm. The thickness tolerance is plus or minus 0.11 mm for sheets over this width. For a thicker sheet, such as a sheet 5 mm thick, the tolerance will be plus or minus 0.19 mm for sheets greater than 1.3 m in width. Some processes of coining and embossing will alter the measurable thickness of the metal. It is the initial thickness before coining or embossing that defines the stainless steel, not the final thickness.

Coils

Sheets begin in the coil form, so lengths are cut to order, but standard lengths are dictated by the cutting and forming operations to follow and the quality limitations of the end product.

For example, most fabrication facilities have equipment established to cut, form, and process lengths that are a maximum of 3660 mm, or 12 ft. Some are equipped to handle larger sheets, but the average press, waterjet, shear, and laser have limits based on the economics of making the equipment and handling the material itself. Stainless steel, because of its superior surface quality, should not be dragged across a bed of steel or lifted in such a way to induce a kink in the metal. A kink is a permanent buckle induced into the metal surface from mishandling. It cannot be removed because the metal has plastically deformed in this local area. Limitations of size come into play, not simply for economic reasons, but for practical reasons. In addition, shipping skids for transferring sheets cut to length are aligned to these lengths as well. Otherwise, larger, specialized skids requiring specialized handling equipment will need to be utilized. Simply joining two skids together to make a larger skid can induce distortion into the metal due to uneven or flexible skid supports.

One other parameter that plays an important role in the length of a sheet is flatness. Stainless steels hold residual stress from the coiling operation. The process of making the thin ribbon under pressure and coiling induces variable stresses in the ribbon of metal. When removed from the coil, the sheets undergo a stress reduction process called a leveling line, which reduces these stresses by distributing them across the sheet; this flattens the sheet.

You can obtain coils of metal that are several hundred feet in length, but the metal will need some post-treatment to make it useful for architectural or artistic use. A modern decoiling line will always incorporate a leveling line in the process. These lines are programmable to cut the material to a specific length, remove the internal stresses in the sheet, coat one side with a protective film, and stack the sheets onto skids for storage and shipping.

Coil processing is an efficient method of storage, handling, and transport of stainless steel sheet both at the Mill and at secondary processors. The coils can be slit to narrower dimensions and recoiled. Coils can be cut to length or be engaged into a stamping or blanking line, thus reducing waste and improving handling. Often, large master coils are produced at the Mill and shipped to be uncoiled and recoiled into smaller coils for easier handling and storage.

The process used on stainless steel coils involves creating a master coil at the Mill. This is a large ribbon of metal, often weighing more than 5000 kg. This coil will be uncoiled and have the edges trimmed to aid in registration as it is further processed.

The master coil goes to an uncoiler, which pays out the metal ribbon to a tensioning device, which in turn removes the crown in the ribbon of metal. The center of the ribbon on a master has a slight crowning that can be set into the metal and must be tensioned for removal, which improves flatness. Leveling lines may be used as well. These differ from tensioning in the way they stretch the metal to correct wavy edges or center distortions. Often decoiled ribbons of metal have edge wave due to the fact that the edges are slightly longer than the center of the ribbon. Distortions in the center occur if the center portion is slightly longer than the outer edges.

As the master coil is decoiled, tensioned, and leveled, it travels to a slitting line to be slit down to narrow coils, decoiled and recoiled into smaller coils, or cut to length and skidded into sheets. This is done at the Mill or at a secondary plant that handles slitting and decoiling processes.

Modern systems incorporate some quality inspection criteria to identify coil breaks and chatter that can be induced into the ribbon of metal as it is processed. Coil breaks should be culled from the line and recycled. This should happen at the Mill. These are conditions where there is a mark running across the width of the metal. They appear as parallel lines running the width of the ribbon of metal. Coil breaks develop in a ribbon of metal as it cools. The metal undergoes localized annealing and as it is uncoiled and then recoiled this creates stretching in the ribbon of metal. A distortion across the metal is imparted.

These coil breaks need to be identified or the coil will be covered with protective coating and shipped like the rest of the metal. The fabrication plant may not see the coil break because it is covered by the coating, and further operations on the sheet then may be performed along with the balance of the metal, only to later peel the protective coating off and uncover the malady. The expense associated with the break will have increased due to all the finish work that has been performed. There is no good way to conceal coil breaks, and subsequent polishing operations will make them more pronounced (Figure 5.5).

Other maladies that can manifest in sheet material are chatter marks. These are lines (sometimes very subtle lines) that cross the sheet. They are induced in the coiling and decoiling process where force has not been consistently applied and the metal has been tugged, relaxed, and tugged as it is uncoiled. Adjustments to tensioning of the ribbon of metal may not be consistent from one end of the coil to the other as the diameter of the coil decreases. Chatter marks can also occur when finishes are applied to the coil material. Any change in tension can create localized polishing inconsistencies.

Metal with coil breaks and excessive chatter marks should not be acceptable material for art and architecture projects. Quality assurance practices at the Mill and at then again at the fabrication facility should be able to cull these maladies from ever making it to the finish product. The difficulty lies in the protective film. The film conceals many of these conditions and further costs can go into a stainless steel fabrication before the malady is uncovered. There is no repair process that can overcome this condition if it is present.

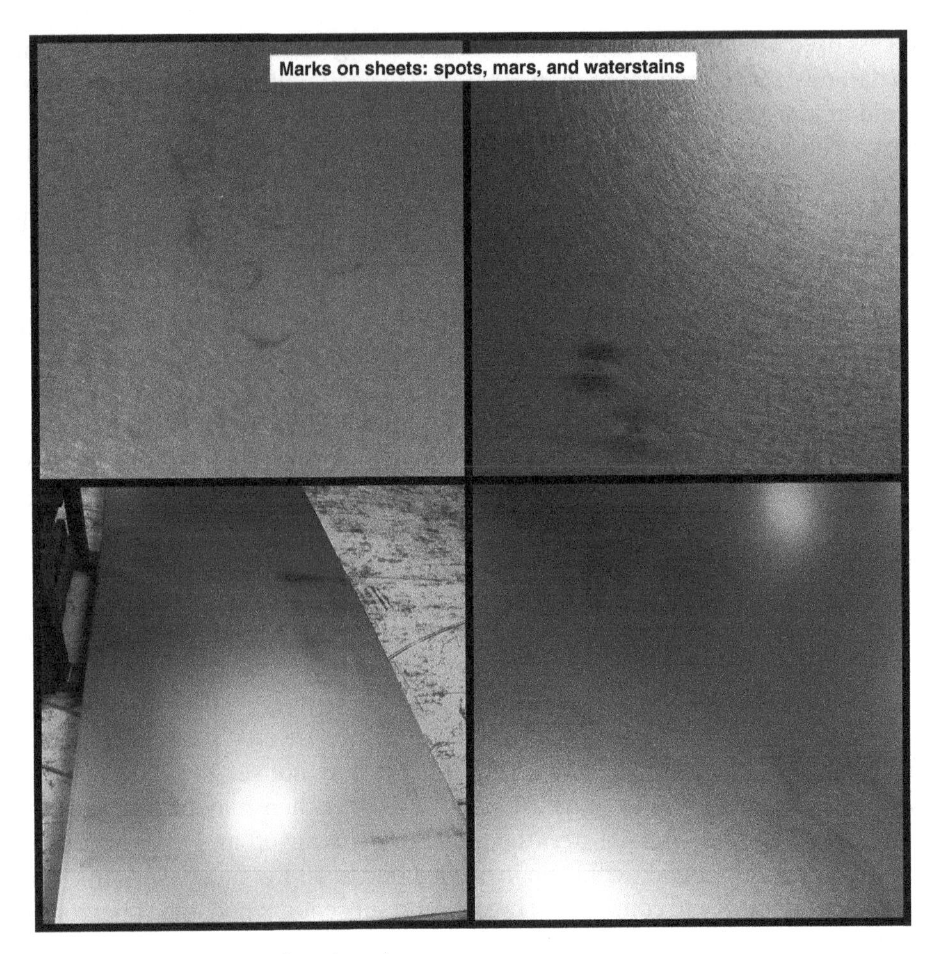

FIGURE 5.5 Various imperfections found on sheets.

Some of these conditions are apparent on the reverse side of the sheet but this again is a problem because a fabrication facility works with the finish side up. It is the point where the protective coating is applied that these imperfections can be caught and not passed on otherwise unnecessary costs will be incurred. A fabricator should have a relationship with the Mill supplier to insure an adequate quality procedure is followed and understood by all parties (Figure 5.6).

The widths of sheets are the other limitation. The widths are limited by the Mill's ability to roll and apply pressures to reduce the thickness from the hot form to the thin form. These limitations are physical constraints. Significant pressure is applied by a mill with small-diameter rolls called a Sendzimir mill or Z-mill. The Sendzimir mill has 20 larger rolls applying pressure to a set of small

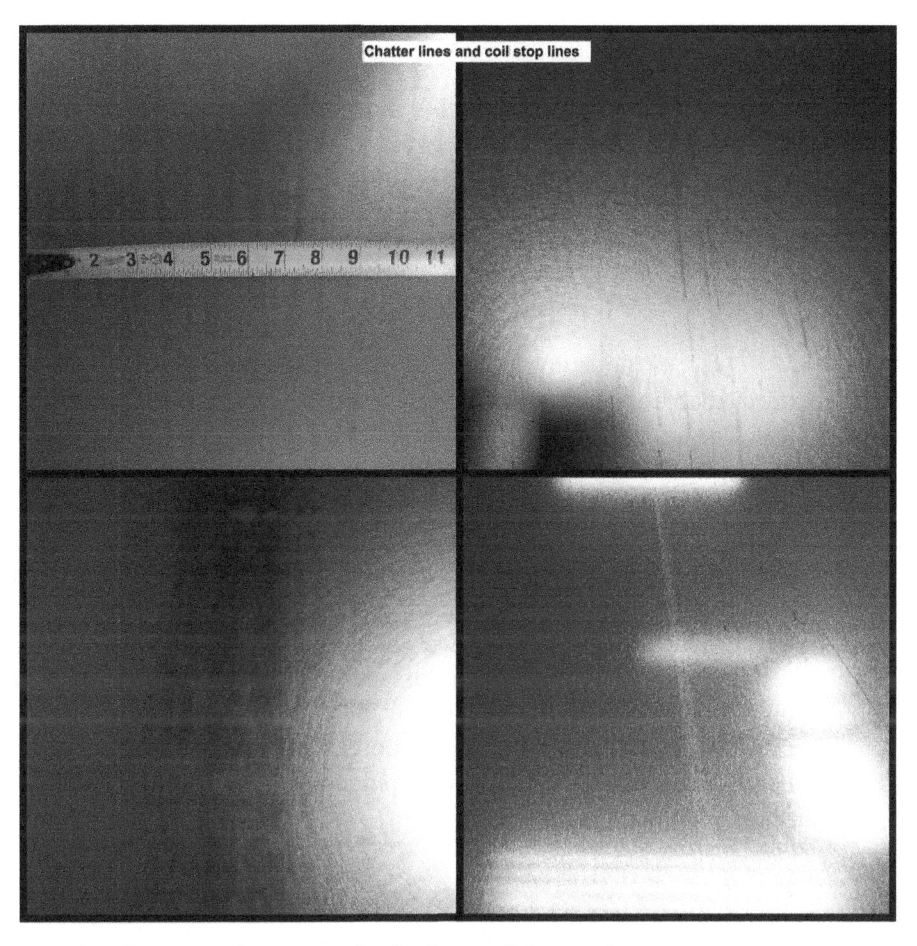

FIGURE 5.6 Various lines sometimes appearing in sheet stainless steel.

rolls 40 mm in diameter in order to squeeze the ribbon to a specific thickness. The longer the roll, the more difficult to apply even pressure and arrive at consistent stress. Widths of stainless steel are limited by production processes and ultimately by the Z-mill.

All subsequent handling and processing operations are designed around these limitations. Coiling and decoiling, flattening, packaging, and laser cutting beds are all designed to accommodate most of these widths. Custom slitting operations can take the standard widths of a coil and slit them down to requirements of the project or fabrication facility. Pricing is higher because the off fall is recycled as scrap. You will be charged for the original coil width so the unit price on off-size coils will be higher.

Typical Widths

Millimeters	Inches
2000	78
1524	60
1500	59
1219	48
1000	39
914	36

All the finishes are available on sheet material. They were developed for application on the sheet forms of stainless steel. The art and architecture marketplace demanded more specialized surface quality and this enabled innovation in finishing techniques. The mill finishes, the satin directional and non-directional, bead blast, embossed and coined finishes, and chemical finishes were created for the architectural marketplace and are available on all sheet material forms of stainless steel. The Mill does not provide these finishes but secondary finishing operations will. There may be some minimum quantity and availability restrictions that come into play, so it is advisable to involve the manufacture of the end production to establish levels of quality and availability.

On coils, only the directional satin finishes, embossed finishes, and coined finishes are currently available. Application of other finishes will become available as process improvements overcome the limitations of applying these finishes on a continuous basis. In the past several decades, new innovations in finishes for stainless steel sheet have exploded, improving on the design possibilities available to the artist and architect.

PERFORATED AND EXPANDED STAINLESS STEEL

Stainless steel sheet is a perfect base material for perforation or expanded metal. Perforated and expanded metals are exposed to the atmosphere on all sides and along the perforated and cut edges. The corrosion resistant aspect of stainless steel makes this an ideal material due to the fact that performance of the protective oxide layer extends to the exposed edge as well.

Perforated Stainless Steel

Perforated stainless steel is used as screening material and as decorative panels. All sheet forms and finishes can be perforated and custom perforated. There are some limitations as to the width of perforated stainless steel. The overall width is dependent on the width of the machine that produces the perforation by punching out holes and shapes. Length is less an issue, however, there exists some practical limitations of handling and subsequent forming operations.

It is important to note that in sheet metals one side is considered "prime," or the face side, while the back side of the sheet does not receive the same level of care. On perforated surfaces the back side will be less ideal and may require some additional finishing if it is to be exposed to view. Often the process of perforating can scratch the reverse side as the sheet moves across the bed or is passed through a set of piercing rolls. You can protect the reverse side with a PVC layer, but this will add cost both in terms of the protective film itself and in the time needed to remove the film (Figure 5.7).

FIGURE 5.7 Operable perforated metal skin on the Cooper Union in New York, designed by Morphosis.

Another challenge with perforated stainless steel is that when the sheet is pierced, the metal work hardens at the edges of these piercings and can induce localized stress into the sheets, causing them to warp or shape slightly. This phenomenon can slow the processing down so as to prevent damage as a result of the sheet catching a slug. The slug is the term given to the waste part pushed out as the process occurs. If slugs get caught in the tool or wedged in the part, they can damage the work. Subsequent flattening operations may be necessary to bring a warped sheet back to flat.

Sharp tools are critical for piercing stainless steels. Dull tools induce stress into the stainless steel, as they tend to rip the edge rather than pierce it with a clean shear. Treated and coated punching tools help reduce heat during the piercing operation.

When perforating stainless steel there are several important relationships to keep in mind. For round holes, the ratio of the diameter to thickness is critical. A ratio of 2:1 for stainless steel will reduce the chance of inducing stress into the sheet and reduce stress on the tool used to punch. For other shapes, this ratio would correspond to the smallest of the dimensions. For example, if the shape is a rectangle, the ratio would be 2:1 for the shortest distance to the thickness of the sheet (Figure 5.8).

The clearance and die design are critical as well. The tool and die must be designed for the correct clearance of the stainless steel thickness. The die should be designed with rounded corners instead of sharp corners when piercing stainless steel. Sharp corners will damage the die and can create small rips in the corner of the piecing (Figure 5.9).

Perforated metal is extensively used in architecture and art. Stainless steel is the metal of choice because of the consistency throughout the metal and the strength afforded. Perforations using

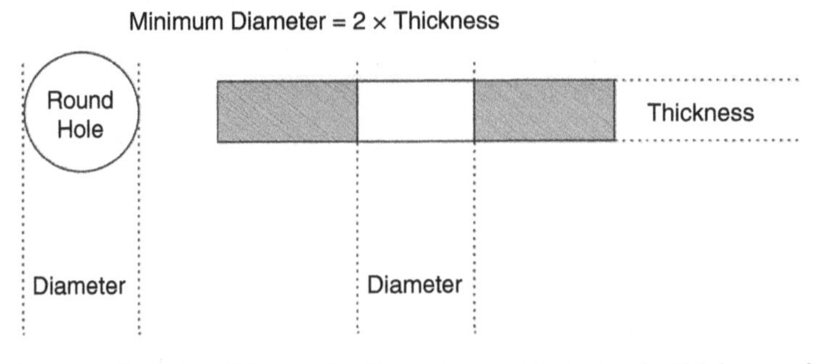

FIGURE 5.8 The minimum diameter of the perforation hole must be twice the thickness of the sheet.

Eliminate Sharp Corners

Rounded Corners Preferred

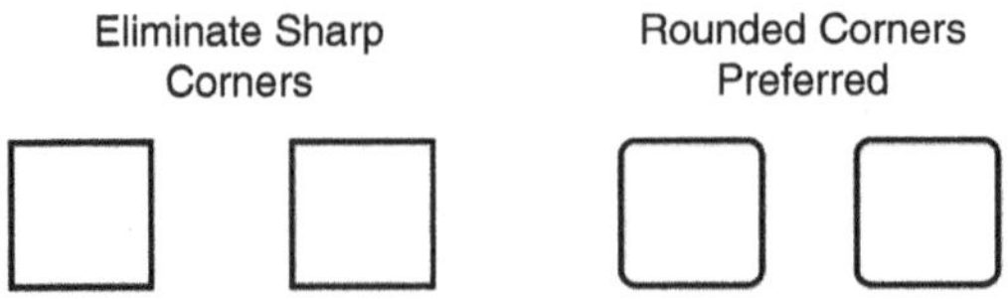

FIGURE 5.9 Eliminate sharp corners in the die.

FIGURE 5.10 The Dream Hotel in New York, designed by Handel Architects.

punching operations often are symmetrical: circles, squares, rectangles, and even stars and shell shapes. Perforations do not have to be symmetrical, but nonsymmetrical shapes can get caught in the tooling. Cutting perforations with a laser affords an infinite variety of shapes. Here again, nonsymmetrical forms and even symmetrical forms smaller than the slats used to hold the stainless steel can twist and interfere, slowing the process down (Figure 5.10).

Expanded Metal

Stainless steel is well suited for expanded metal. Expanded metal involves slitting the stainless steel at preestablished intervals and then either upsetting the area at the slit to cause one side to extend

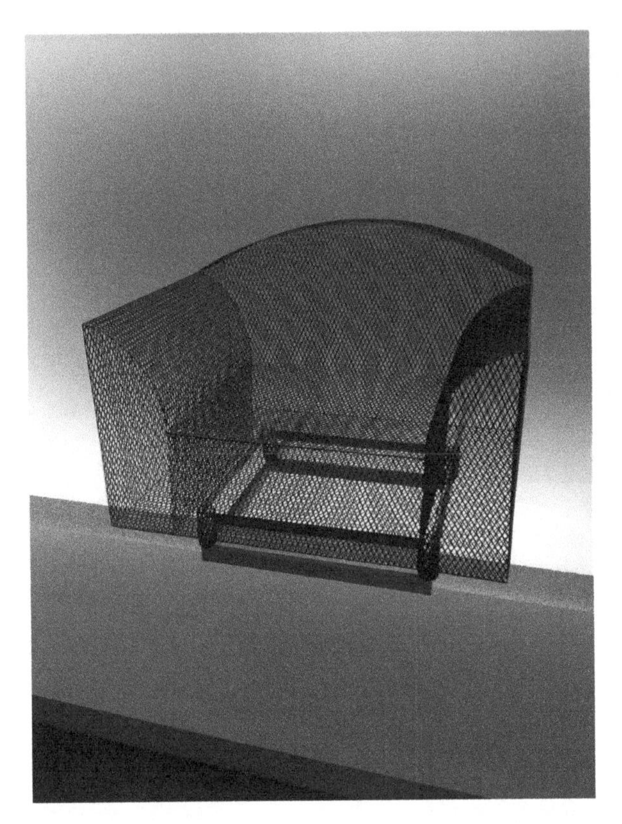

FIGURE 5.11 A chair made from expanded metal.

out and the other side to extend inward or pulling and stretching the metal to open the slit into a diamond-like perforation. No metal is removed in the process, and by expanding the material better coverage can be obtained. This allows for a better yield of metal coverage. As the metal is slit and stretched, the overall area increases drastically.

There are various sizes and configurations of expanded metal. The typical finish available is a mill finish, but some supply houses will pierce special finishes. Typically, the openings made in the expanded metal have a slightly raised "lip," which is created as the stretching occurs. This form is known as standard expanded metal or formed expanded metal. This formed sheet can be further processed to flatten out the lip, and is referred to as flattened expanded metal (Figure 5.11).

Expanded stainless steel sheet can be ordered with a diamond pattern running lengthwise or across the sheet. The diamond pattern can be staggered or straight line. The diamond can be created in a multitude of sizes, from very small on thin sheet to very large.

Other decorative patterns are also available from different manufacturers of expanded metal. The availability of sizes and thicknesses of more custom openings should be investigated directly with the manufacturer (Figure 5.12).

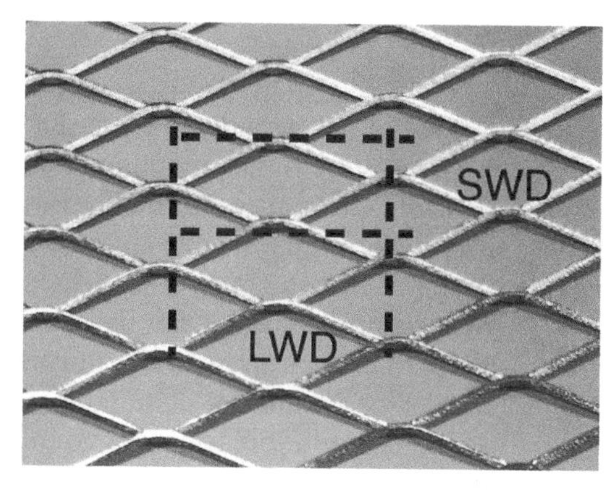

FIGURE 5.12 Typical expanded metal.

FIGURE 5.13 Typical (*left*) and micro-expanded (*right*) metal.

Expanded metal is usually ordered by the "long way of the diamond" (LWD) and the "short way of the diamond" (SWD), as shown in Figure 5.12. The width of the strand, or metal connector, is also specified, along with the metal thickness. Expanded metal comes in many sizes and dimensions. Even very small mesh sizes are possible with thin stainless steel to create screen coverings (Figure 5.13).

PLATE

Stainless steel thicker than 4.76 mm (0.187 in.) is considered plate. The main distinction between stainless steel plate and stainless steel sheet is the finish and refinement of the surface (Table 5.2). Sheet stainless is passed through the Z-mill to reduce its thickness, and subsequent passes are through cold rolls. This elongates the grains and smooths the surface.

The plate form of stainless steel may see the planetary rolls, which reduce its thickness. Plate is rolled hot, so the finish surface is not as refined. Plate can have inclusions on the surface and waviness across the surface. Thin plate can receive most of the finishes applied to sheet and the surface quality will improve. But as the thickness of the plate increases, the surface of the plate becomes less flat, and there will often be pits and inclusions apparent on the surface. There is a drop-off in consistent quality due to the inclusions that might be present in the casting of the metal and the subsequent rolling while hot.

Plate stainless steel is available in several standard forms. Plate width dimensions are similar to those for sheet. Depending on the source, plates of various thicknesses are available in widths up to 3 m (120 in.). Standard widths are shown in Table 5.3. These are nominal, with dimensions of plate from European and Asian suppliers rounded to the nearest millimeter.

Plate lengths are limited by handling and transport considerations. Plates cannot be leveled in the ways that sheets can. They can be rolled to even out the stress in thinner sections if the rolling facility can handle the weight of large stainless steel plates. Plates are not as flat, as sheet forms of stainless steel and the surface will have variations due to cooling from the hot form. Plates passed through a cold roll pass have a smoother, more refined surface. In all cases, though, architectural surface finishes produced for plate forms of stainless steel are not equal to the finish levels obtainable in sheet forms of stainless steel (Table 5.4).

TABLE 5.2 Finishes on plates.

Finish form	Description
Hot rolled	Scale on the surface: very rough
Hot rolled and pickled	Scale removed: rough surface
Hot rolled, pickled, and temper passed	More refined surface: smoother
Hot rolled, pickled, and cold rolled	Improved surface with fewer inclusions
Polished	The best surface available on plate; still not as good as sheet

TABLE 5.3 Standard widths of stainless steel plates.

Millimeters	914	1000	1219	1524	1829	2438	3048
Inches	36	39	48	60	72	96	120

TABLE 5.4 Plate thicknesses.

Millimeters	Inches
4.8	0.188
6.4	0.250
8.0	0.313
9.5	0.375
11.1	0.438
12.7	0.500
14.3	0.563
15.9	0.625
22.2	0.875
25.4	1.000
31.8	1.125
38.1	1.500
44.5	1.750
50.8	2.000
63.5	2.500
76.2	3.000
82.6	3.250
88.9	3.500
95.3	3.750
101.6	4.000

BAR AND ROD

Rod and bar forms of stainless steel are similar to those forms of steel. These are solid, stainless steel alloys and are produced in stock lengths by hot rolling, forging, or extruding. Maximum lengths are what can be handled and transported. For bar material, the thickness is 3 mm (0.125 in.) or greater, with widths to 254 mm (10 in.). Widths in excess of this are considered miniplates. Rod material is available in rounds (circular cross-section), octagons, and hexagons. Minimum dimensions in diameter are 3 mm (0.125 in.).

The mill finish is a dull, pickled, scale-free surface. For artistic and architectural uses, the bar material goes through cold finishing processes. The surface can be finished to satin, and even mirror, finishes. The satin finish is linear: along the length of the piece. For rounds, a satin finish can be produced around the circumference.

Bar material goes through a process known as "turning" to remove defects on the hot rolled surface prior to further finish operations. Turning is a milling operation that cuts away the surface until a specific dimension is arrived at. Another process used is grinding, also called centerless grinding, where the bar material surface is ground down to a certain dimension. This arrives at a surface better suited for further finishing.

There are many variations on standard bar and rod that are produced by cold drawing through a die. These are limited in dimension and require significant quantity to account for the die cost, but they reduce machining time and provide an excellent surface.

Other custom shapes can be hot rolled. The finish is tougher and requires subsequent pickling to remove scale. Hot rolling is used to produce long straight forms that will undergo subsequent welding, shaping, or forging operations (Figures 5.14 and 5.15).

FIGURE 5.14 Welded stainless steel rod sculpture, designed by Caleb Bowman.

FIGURE 5.15 Examples of rod stock.

COLD DRAWN STAINLESS STEEL SHAPES

Cold drawn stainless steel is a precision reduction process used to develop a linear shape with a precise cross-section. A bar or rod is passed through a series of dies that shape the metal into a precise cross-section. The benefits are a very accurate cross-section with very good surface qualities. The process produces near net shapes with little waste.

The cross-section and the quantity minimums are the limiting factors. The shape needs to be less than 75 mm (3 in.) in diameter and not overly complex. The process needs a minimum quantity, which is dependent on the cross-section configuration. Some of the smaller angle and channel shapes shown in Figure 5.15 are drawn shapes.

TUBING AND PIPE

Stainless steel piping is produced in significant quantities for the oil, gas, and chemical processing industries. Pipe is distinguished from tube in that pipes are vessels for transporting fluids and the inside dimension is critical for what is being transported. The outside diameter and pipe thickness are nominal values. Pipe is provided from the Mill with a hot rolled and pickled surface. No particular care is extended to this normally industrial surface beyond removing scale and providing a clean passive surface.

Tubes, on the other hand, are structural and the outside dimension and wall thickness are exact values. Tubes are cold worked to achieve accurate dimensioning. They can be seamless, produced by hot rolling processes with initial surfaces similar to hot rolled surfaces requiring further finishing. Tubes can also be created from hot rolled plate and cold rolled sheet, and are strip shaped in rollers into the tube form and welded along the length. The weld is ground down and prepared to reduce appearance effects. This occurs on both the inside and outside surfaces. The allowable sulfur is usually higher in these alloys to facilitate high-speed welding operations.

Architectural uses of stainless steel tubes are incorporated into the exposed structure when there is a need for a corrosion resistant element that possesses strength and appearance characteristics (Figure 5.16). Pipe is always round in cross-section, while tubing can be round or rectangular in cross-section.

Tubing can be finished by most mechanical means. For satin finishes, it is best to describe the grit size of the finish rather than specifying a No. 3 or No. 4 finish. Non-directional finishes as well as mirror polishing can be applied to tube surface. The welded seam, however, may show when the surface is mirror polished or chemically colored. Tests on the finish should be performed to arrive at an acceptable appearance.

FIGURE 5.16 Hope Tower at University of Nebraska Medical Center is made of stainless steel tube and physical vapor deposition gold custom perforation and was designed by James Carpenter.

WIRE, WOVEN WIRE, AND SCREENS

Stainless steel is available in multiple wire forms, both single strand and multiple strand. Single wire is available in diameters 0.254 mm (0.01 in.) up to 12.7 mm (0.5 in.). Multistrand cable can be developed from wire into massive assemblies. Stainless steel wire is provided in a semifinished state free of scale. Wire is produced from rod material and cold drawn to specific dimensions and structural characteristics and tempers, from spring temper to fully annealed.

Wire screens made from weaving metal threads on special looms similarly to woven cloth is available in a multitude of forms. Simple screening is available in several sizes called sieve sizes. These can be woven to very tight tolerances. The stainless steel gives these screens strength and corrosion resistance (Figure 5.17).

Decorative woven stainless steel screens are produced by a number of companies that specialize in woven wire. Many of these wires started in a more mundane industrial context, as belting for conveyor systems. Woven stainless steel wire can be almost cloth-like in terms of its fine texture. The diameter of the wire, as well as the temper, determines the flexibility and strength of the woven wire screen.

Woven stainless steel is very strong and durable. All alloys of stainless steel are woven into mesh screens. These screens need to be tensioned and spring loaded if the design intention is to keep them looking taut. The stainless steel will move thermally and shift, and the tensioning will keep the surface looking taught and flat.

FIGURE 5.17 Woven wire screens.

Woven wire screens are made from round wire, semiround wire, and flattened wire. The wire can have diameters as small as 0.02 mm. Various sizes and shapes can be intermixed. Stainless steel wire weaving can be incorporated with other metal wires, such as those made of brass or copper, or with other materials, such as Kevlar® and nylon, to create decorative strong screens. Post-processes, in which the tops of the weave are polished or the wire is colored by heat or interference processes, can add to the versatility (Figure 5.18).

Variables available are numerous, and as a design material woven stainless steel wire has tremendous potential. The post–surface treatment has several opportunities to enhance appearance: polishing the tops, electropolishing, heat tinting, and even interference coloring.

Variables to consider are:

- Wire size
- Wire finish and type
- Weave aperture
- Weave type
- Surface treatment and polish

Stainless steel wire is used as support for many hanging artistic elements. Stranded wire provides superior strength and resilience. It comes in a number of sizes and there are custom fittings forged from stainless steel bar and rod stock made specifically for use with stainless steel cable. Stainless

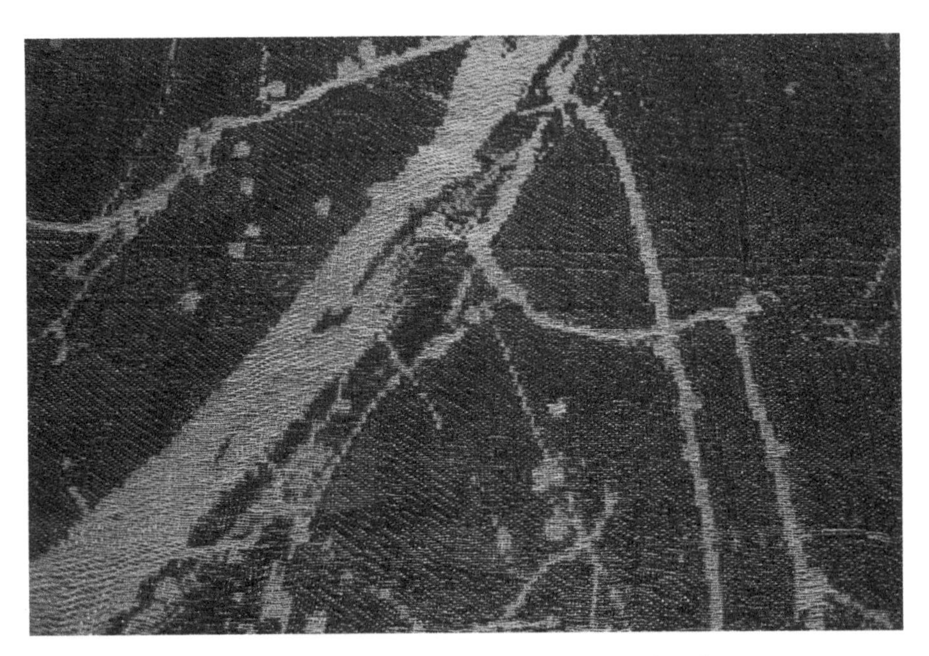

FIGURE 5.18 Woven wire art prototype made from stainless steel wire and Kevlar.

FIGURE 5.19 Hanging glass art utilizing stainless steel cables, designed by Dierk Van Keppel.

steel wire and cabling can be used with specially engineered stainless steel fittings to give tensioned support systems a minimal yet elegant solution (Figure 5.19).

Stainless steel cable and wire come in various sizes and rated load capacities. The typical alloys is S30400, but other alloys are available through special order. Sometimes referred to as aircraft cable, stainless steel wire comes in several diameters and load capacity ratings. Fittings can be swedged onto the ends for special attachments and decorative appearances.

EXTRUSION

Stainless steel can be extruded. The shapes possible are limited in outside dimension because they need to fit within a certain diameter circle, typically no more than 165 mm (6.5 in.). The process

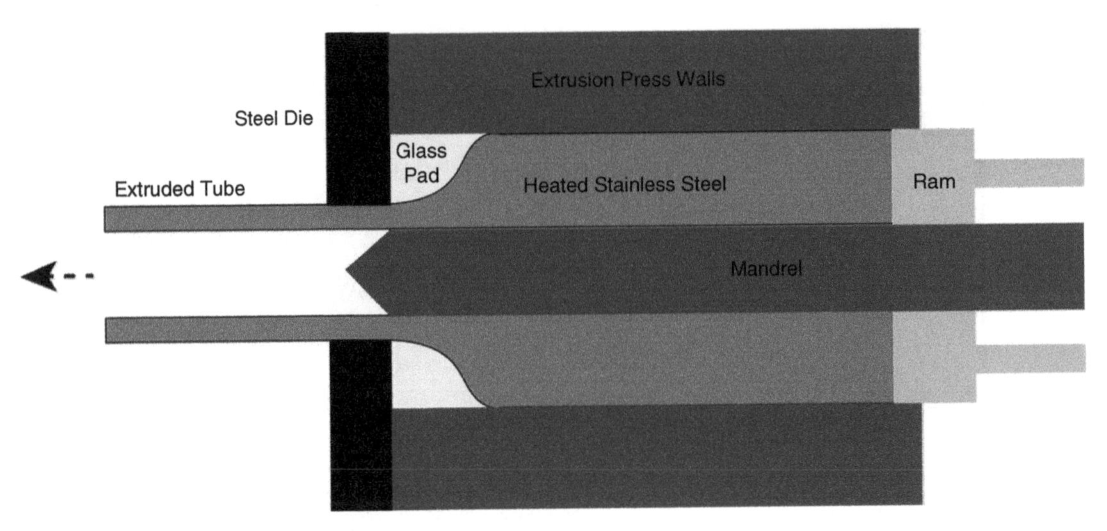

FIGURE 5.20 Process of extruding a hollow stainless steel form.

of extruding metals goes back centuries, but it wasn't until the early 1950s that stainless steel extrusion was introduced commercially. The process used to extrude stainless steel is called the Ugine-Sejournet process. It involves using glass as a lubricant and an isolator of the hot stainless steel billet. In the process, the billet of stainless steel is heated to around 1200°C, slightly higher than the temperatures used for hot rolling. The hot billet of stainless steel is rolled in glass powder that melts and adheres to the outside surface. A glass pad is placed toward the back face of the extrusion die to insulate the die from the hot stainless steel.

Extruding of seamless stainless steel tubes was a major market in the past, but continuous welded tubes are more economical to produce and have replaced extruded tubes in much of the industry. There are some shapes better fitted for extrusion, but the market for stainless steel extrusions is limited (Figure 5.20).

STRUCTURAL SHAPES

Stainless steel structural shapes have become as much an aesthetic choice as a corrosion resistance choice. There has been significant interest in the use of stainless steel for structural elements in both art and architecture. Chosen for the aesthetic appeal and enhanced performance stainless steel offers, the use of stainless steel structural shapes should continue to grow.

The finish provided on stainless steel structural shapes is usually produced by blasting the surface with abrasives. Typical abrasives used are stainless steel shot, glass bead, and crushed quartz. Other finishes, such as Angel Hair, Vibration, and satin brushing, can be used, but they are more expensive and more difficult to apply, requiring more hand application than mechanical or computer controlled finishing.

The most common structural shapes are produced from austenitic alloys of stainless steel. Duplex alloys come into consideration when high strength is desired: section size reductions can be achieved when using the duplex alloys due to their significant strength.

Stainless steel structures come in several forms:

- Cold formed shapes produced by press brake or roll forming
- Hot formed shapes: angles and channels
- Hot formed and welded shapes: I-sections and hollow sections
- Extrusions (limited to fitting within a 330 mm circle)
- Welded plates
- Machined castings or bar (Figure 5.21)

FIGURE 5.21 Stainless steel structural shapes.
Source: Photo courtesy of Stainless Structurals America.

Stainless steel structural shapes are produced by hot rolling processes and by laser fusion of plates to form the shapes. The hot rolled shape requires rounded corners, as the material is shaped by passing a solid linear form through shaping rolls. The metal is heated to just above the recrystallization point of stainless steel to make it pliable and able to be formed into the desired shape. The red-hot form is passed through steel rollers that apply tremendous pressure as the new shape is created from the stainless steel billet.

Laser-fused shapes are made from fused plate sections of stainless steel. The finish on the laser-fused sections is superior and can be finished readily for artistic and architectural use. The corners are square, as the plates are fused together. Angles, beams, channels, and tubes can be made both via the hot rolled method and laser fusing.

Many structural shapes are made from laser fusing flange plates to web plates. W-sections, I-sections, and T-sections can be made with specific structural properties. Angles and channels are typically hot rolled into forms.

Stainless steel fittings, linkages and brackets made from welded plates, milled blocks, and forgings are used extensively in the building construction world. The combination of strength, corrosion resistance, and decorative appearance allow for sleek machined forms that can be used to hold large glass panels in position while at the same time adding design elegance to the wall.

DESIGNING WITH STAINLESS STEEL STRUCTURAL FORMS

Stainless steel has a strong anisotropic nature in the direction of the grains. As the material is rolled out in the processes of manufacturing, the grain direction will play an important part in the tensile and compressive strength in the direction of the grain. Additionally, stainless steel has a higher coefficient of thermal expansion than that of carbon steel, so as much as 30% greater thermal movement can be expected. This is a critical factor if designing stainless steel structural pieces to perform alongside carbon steel members. The stainless steel is going to grow or shrink at a greater rate than its steel cousin.

Accompanying the higher coefficient of expansion is a thermal conductivity lower than that of carbon steels. This is a critical factor to consider when welding sections, because this combination of traits can increase the residual stress developed in welding operations.

Austenitic stainless steels are very resilient. They can absorb stress from impact better than carbon steels, which makes them ideal for explosion- and impact-barrier material. They can absorb impact forces without fracturing due to their strain-hardening behavior.

Stainless steel structural properties also differ from carbon steel properties as depicted in the shape of their stress–strain curves. For stainless steel, the stress–strain curve is more rounded and lacks a well-defined yield stress point, whereas carbon steel exhibits a linear elastic behavior as it approaches yield, then plateaus before continuing to strain hardening. For stainless steel, there is no plateau (Figure 5.22).

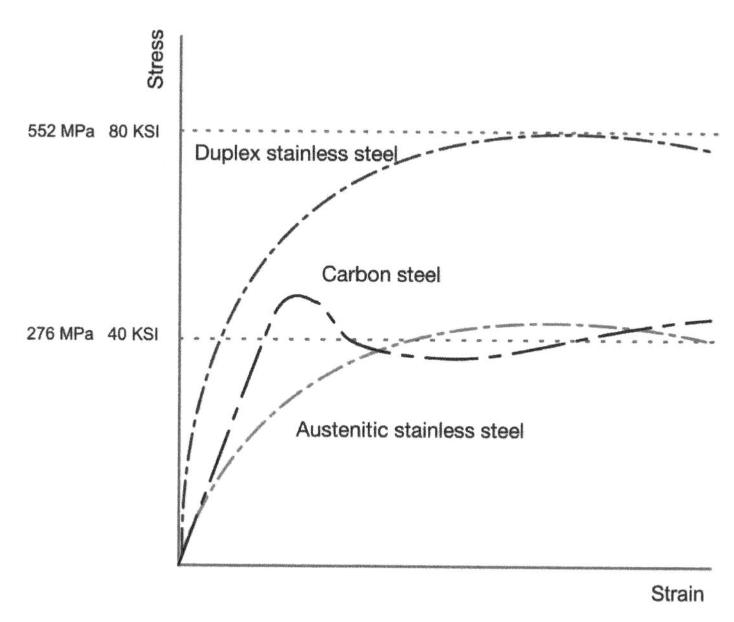

FIGURE 5.22 Stress–strain curve of stainless steel alloys.

When designing with stainless steel structural forms, several differences from carbon steel need to be considered:

- Anisotropy
- Nonlinear stress–strain relationship
- Low proportion limits
- Pronounced response to cold work

Anisotropy refers to a condition where mechanical properties in the longitudinal direction are different from the mechanical properties in the transverse direction as it relates to tension and compressive stress. This condition is characteristic of stainless steels.

The nonlinear stress–strain relationship is flatter for stainless steels than for carbon steels, and the anisotropic aspect is different still. The stress–strain curve is more gradual for stainless steels than for carbon steels, the yield is smoother, and the proportional limits are lower. For carbon steel the proportional limits are a minimum 70% of yield, whereas for stainless steel they are 36–60% of yield. This can affect the buckling behavior predicted and is a design consideration. To counter this, higher tempers are often used.

The factor of safety for stainless steel is 1.86, while for carbon steels it is 1.67. Part of the reason behind the higher safety factor is that there is less experience with stainless steel. A deeper

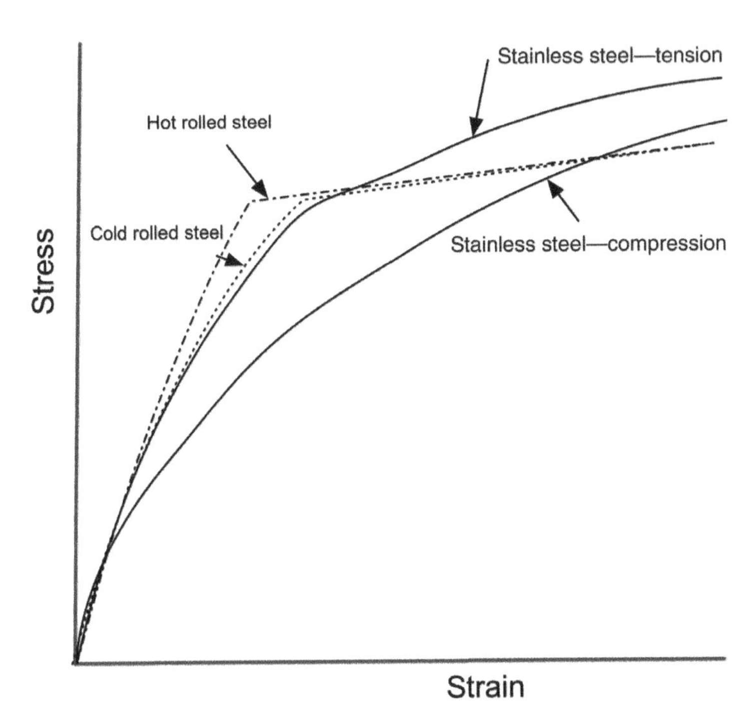

FIGURE 5.23 Stress–strain relationship of steels.

understanding, testing, and history will be needed to allow a reduction of the factor of safety (Figure 5.23).

As stainless steel passes the point considered the proportionality stress point, yielding becomes gradual and this leads to a softening of the stainless steel and a corresponding reduction in buckling resistance. For this reason, the design strength used for stainless steel differs from that of carbon steel structural elements. These mechanical and material property differences are the reasons why there is a different specification for stainless steel structural shapes. The research and analysis on structural stainless steel is ongoing. The desire to use stainless steel structural members as architectural features is a powerful driver of this research and analysis, which will provide a deeper understanding of the performance characteristics of this highly corrosion resistant metal. The long-term savings due to lower maintenance costs, longer life-cycle cost, and the sheer beauty of the metal make stainless steel structural components a desirable material for design.

POWDER, BEAD, AND SHOT

Stainless steel is available in powder form, small beads, and stainless steel shot. These minute forms are available in S30403 and S31603 alloys. Other alloys are available, but only through special request. Stainless steel powders are very small: as small as 300 μm (0.01 in.) in diameter.

Decorative stainless steel beads and stainless steel shot are cast as spheres or cut from wire. They come in various sieve sizes and are used extensively to clean other metals. Used in this way, they do not fracture and do not deposit potential substances that could introduce steel or iron.

Stainless steel plates blasted with stainless steel shot can form a very hard surface. The shot work hardens the surface, making it resistant to scratches and mars.

Beads of stainless steel are available in a number of sizes. Used mainly for jewelry, these cast spheres can be tumbled and polished.

FOIL

Stainless steel can be rolled very thin to foil thicknesses and provided with various tempers. The available material is narrow strip and has a reflective surface finish due to the numerous passes through reducing rolls.

Typical foil alloys are S30400, S30900, and S32100. Stainless steel hardens as it is cold worked, so the process of producing sheet dimensions to thin foils is difficult and requires annealing steps. The edges of these foils are razor thin, making handling and working with thin foils a delicate process. Foils are used as corrosion barriers and reflectors, and in a number of other situations that require a durable nonreacting material. Available foil thicknesses are from 0.127 mm (0.0005 in.) to 0.254 mm (0.010 in.). Foils are often applied to fibrous materials to make them easier to handle. Foils are only available in the No. 2B and No. 2BA finishes. The thinnest of the material precludes the ability to apply other finishes with any consistency.

CAST STAINLESS STEEL

Stainless steel alloys are readily cast in a manner similar to steels. Cast stainless steels are playing more of a role in art and architecture as foundries are seeking new markets and designers are understanding how to incorporate castings into their designs.

The biggest drawback in using castings for art and architecture is their finishing limitations. You can achieve a highly polished cast surface, but at a cost. To achieve a highly polished surfaces limits the casting to being performed in a ceramic shell with very tight alloying controls. This is a more expensive cast method. In addition, to achieve a high polish the surface will need to be carefully sanded using finer grades of grit: down to 1500 grit or finer. The sanding is then followed by polishing with cotton buffing wheels and fine diamond paste to achieve a highly reflective surface.

Large castings are produced in sand cast molds, which results in a coarser surface. Once the cast stainless steel part is removed from the mold, the surface scale is taken down with clean sand or stainless steel shot blasting. Under no circumstances should the use of steel shot or contaminated sand be used to remove scale from stainless steel castings. These will embed particles into the surface and can damage corrosion resistance. The surface of the stainless steel must be very clean and passive before the finishing and polishing of the surface. Once the scale is removed, the surface is pickled, and that step is followed by rinsing and final finishing. Glass bead gives either a fine, light texture or a mechanical brushed finish, such as Angel Hair or Vibration.

The selection of the cast process for stainless steel begins with evaluating these three criteria:

1. Finish surface quality
2. Dimensional tolerances
3. Cost

Each of these factors plays a significant role in determining the value of using cast stainless steel as an option. Finish quality is a function first of the mold used to cast the part and second the skill used in the finishing operation. The desired final finish is another constraint. Producing a mirror polish on large cast parts is very difficult and expensive. The results may not be consistent and may not meet the requirements of the design.

Dimensional tolerance is tied directly to the casting process itself. Small investment castings can have very tight tolerances, as low as 0.1 mm, while large sand casting parts have tolerances

FIGURE 5.24 Cast stainless steel end on a wood column at the Arena Stage in Washington, DC, designed by Bing Thom Architects.

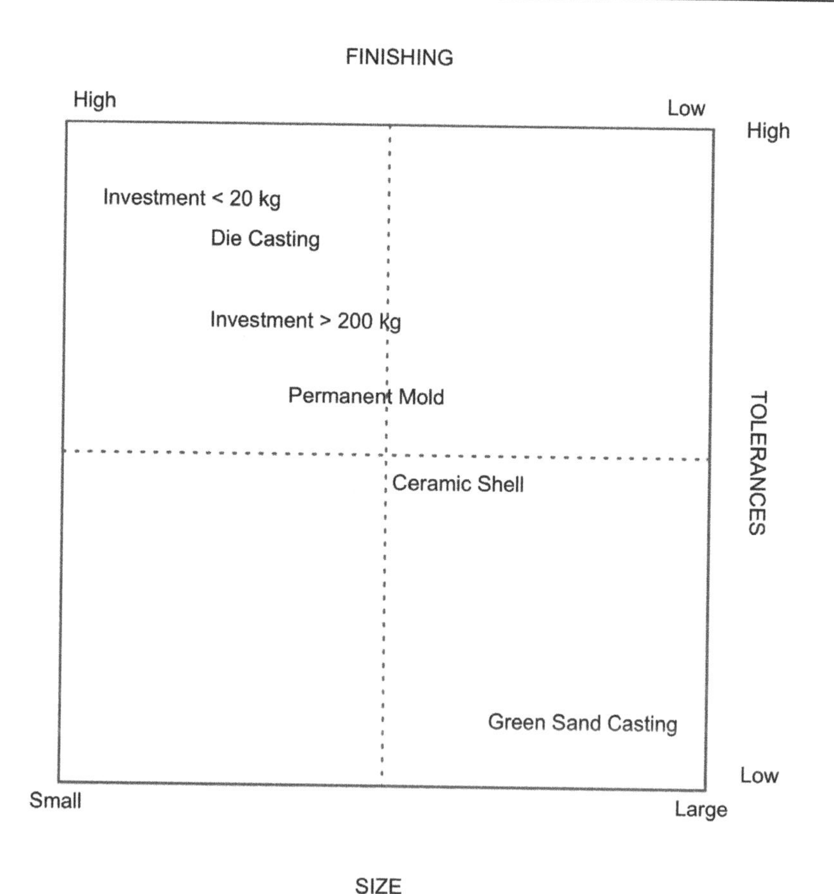

FIGURE 5.25 Matrix of cast processes.

of as much as 12 mm. Machining can aid in achieving tight tolerances, but this will add cost. Cost always plays into design and material choice. In the case of casting, cost is a constraint dependent on quantity, mold process, alloy of stainless steel used, and skill level of the operation doing the casting. Modern processes of producing molds from rapid prototyping equipment coupled with development of ceramic slurry compounds is having a significant effect on reducing time and cost for casting (Figures 5.24 and 5.25).

These are the forms available to the fabricator of art and architecture projects. They are not unlike the forms other metals are produced in. Their size and shape limitations are defined for the most part by industry production processes at the Mill, handling and transportation constraints, and by the equipment used to further fabricate the metal into the shapes and forms of art and architecture.

CHAPTER 6

Fabrication Processes

If you define the problem correctly, you almost have the solution.

Steve Jobs

WORKING WITH STAINLESS STEEL

Stainless steel poses special challenges to the metal fabrication facility. For nearly all art and architecture projects, stainless steel is sought for its reflective surface quality and long-term predictability. Those involved in forming, shaping, stamping, and spinning operations must have a clear understanding of the idiosyncratic behavior of this special alloy of steel. Stainless steel has mechanical and metallurgical properties that make it markedly different from steel. Firms that are versed in fabricating steels will require modifications to their processes when it comes to working with stainless steels.

There are several distinctive characteristics to consider when working with stainless steel:

- Reflectivity: small variations will be enhanced with stainless steel
- Anisotropy: in sheets and plate forms, a directional bias exists
- Nonlinear stress–strain relationship: unlike steel, less abrupt yield condition
- Low proportional limitations: lower buckling resistance than steels
- Rapid cold working behavior: cold working rapidly increases temper
- Springback behavior: internal stresses and material strength
- Thermal conductivity: lower than other metals, does not move heat away
- Corrosion resistance: the surface must be kept passive and clean

Stainless steels have different mechanical properties along the grain direction of a sheet as opposed to across the grain of the sheet. This comes into play when forming edges with the grain versus forming the grain direction of a sheet transversely. The springback behavior of the metal will also be different and require some minor accommodation as you bend the metal.

For example, when forming a pan into a stainless steel sheet, the brake operation will often require adjusting for the springback in the folded edge as perpendicular bends are imparted. More importantly, though, is that the edge must be true and free of the most minor crease. The high reflectivity of stainless steel will enhance the most minor distortions (Figure 6.1).

Nonlinear stress–strain behavior is more of a structural consideration. Austenitic stainless steel will demonstrate inelastic behavior at lower stress levels. This added stiffness aids in using thinner sections. There is a trade-off in terms of internal stresses, though. The thinner stainless steel often has greater internal stresses that can lead to distortion in reflectivity.

When you consider stainless steel elements, particularly large flat elements, the width-to-thickness ratio should be reduced to overcome internal distortions in the sheet. Surface reflectivity will play a part in how the distortion will be muted or accentuated as the viewer moves around the fabricated part.

FIGURE 6.1 Mirror-like surface of a 2BA finish metal roof showing distortions at clip placements. Designed by HNTB Architecture.

TEMPORARY PROTECTIVE COATINGS

Working with stainless steel to create the forms of art and surfaces of architecture demand special care and optimized equipment. Most of the raw material will arrive protected with a thin plastic film adhering to the surface. On flat sheet and plates, the film is often low-tack plastic suitable for laser cutting operations. Additionally, for sheet and plate material it is important that a direction indicator arrow be printed on the plastic film to indicate the direction the material came off the coil. There is a slight anisotropic behavior to the grains of wrought stainless steel. Reflectivity is conditioned differently in one direction versus the opposing direction. This is because as the sheet or plate is rolled, the grains align in the direction of rolling. Very slight color differences and reflective tones can be apparent in the metal. This is less of an issue when mill surfaces are used. Subsequent finishing processes can induce more enhanced directionality in the reflective appearance.

Bars will also have film protecting the surface if a finish has been applied. Tubing and pipe usually are provided without film. Raw stainless steel material is supplied to the fabricator of architecture or art forms with higher standards of protection because their end use has ascetic requirements. Once the material is at the fabrication facility, this added protection must be maintained.

Steel particles from operations on steel parts can contaminate stainless steel surfaces through either contact or transfer through the tooling. Separation of activities is a requirement for a facility involved with stainless steel fabrication. Contamination from other steel operations is a common cause of embedding steel in stainless steel surfaces. This can lead to premature corrosion of the metal.

Tooling and equipment must be robust and accurate. Stainless steel fabricated panels and assemblies demand accuracy. Flaws are more apparent. The edge of stainless steel formed parts is more defined, and distortions induced from fabrication can ruin a finish part. When stainless steel is formed, the edges are sharper and more defined than for other formed metals. Nicks in dies or spaced built-up tooling can impart tiny flaws into the bend. These will show in the finished part due to the reflective nature of the metal.

All the various grades of stainless steels used in art and architecture possess properties of strength and hardness that exceed those of cold rolled steels. All the grades exhibit work hardening behavior that exceeds that of cold rolled steels. In particular, the austenitic stainless steels, which are the more common architectural alloys, have significant increases in strength and hardness from cold working processes. Fabrication processes must be aware of and overcome springback due to the higher yield stainless steels possess. When forming duplex alloys, springback is a significant concern. Not only are the duplex alloys harder with a greater yield strength, they will require compensation for the inelastic character these alloys possess.

The most important criteria of working with stainless steel is protecting the surface from contamination, using clean tooling, setting the right clearances, and maintaining the corrosion resistance qualities of the metal.

CUTTING STAINLESS STEEL

Shearing and blanking of stainless steels involves slightly more power in the cutting shear because of the higher shear strength of stainless steels. Additionally, shearing stainless steel demands addressing the shear blade more frequently to keep it aligned and sharpened. It wears out faster due to the harder metal.

Shearing is an operation involving two sharpened and hardened steel blades. One blade is fixed, and the other is brought down with such force that a metal strip undergoes severe plastic deformation to the point that it fractures at the surface line where it contacts the shear blades. This fracture propagates through the metal. When there is proper clearance between the cutting blades, the crack that propagates penetrates only a portion of the thickness. One crack initiates from the top blade and another crack initiates from the bottom blade. These cracks meet near the middle to provide a clean fracture line. Stainless steel is harder than steels. Signs of wear are rough edges on the sheared stainless steel and a slight roll-over of the sheared edge. This edge can be sharp and jagged due to ripping caused by the shearing blades being out of alignment. Clearance is critical when working with stainless steels. Blanking, punching, and shearing operations on stainless steel can require an increase in clearance of as much as 10% of the thickness of the metal (Figure 6.2).

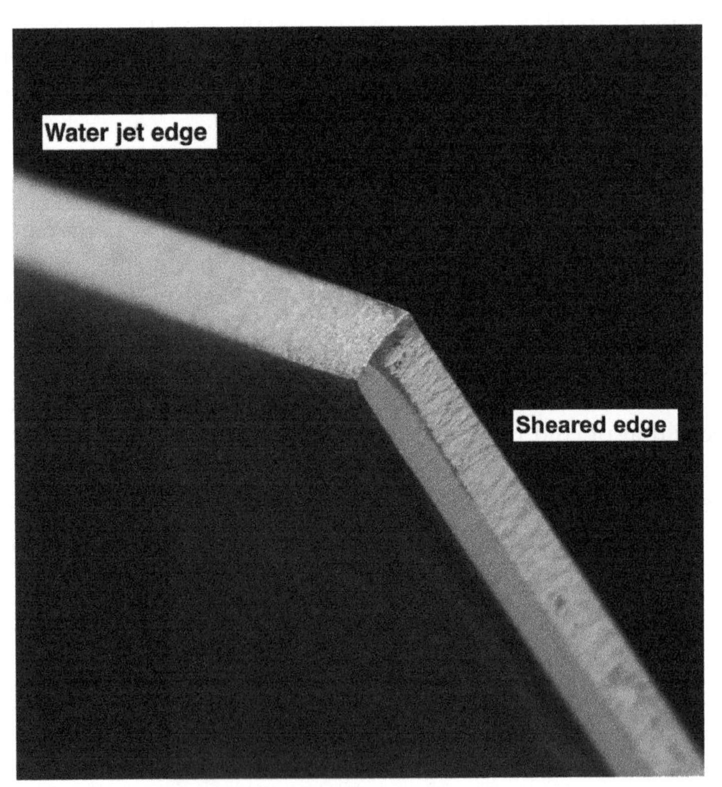

FIGURE 6.2 Close-up of the sheared edge of a 3 mm sheet.

Saw Cutting

Wrought and cast stainless steel can be cut with cold saw blades. Stainless steel is more difficult to cut because of its hardness and because it does not dissipate the heat generated during the cutting process. A cold saw is a steel circular or band saw. The tips of the blades can be coated in tungsten carbide and coated with synthetic diamond or with titanium nitride. Saw cutting can induce heat into the stainless steel, causing localized distortions along the edge. It is good practice to conceal a saw cut edge in a cover plate or by folding it under to create a lap joint. It is very difficult to achieve an acceptable saw cut edge in thin stainless steel sheet. Bar, tubing, and piping can be saw cut with acceptable edges along the cut due to the greater section and thicker material.

Laser

As they are for steel, lasers are excellent and efficient cutting tools for cutting stainless steel alloys. Intricate shapes and detail can be cut into stainless steel surfaces. A laser works by concentrating a highly focused beam of light onto the metal surface. The electromagnetic energy of the laser light is transformed into thermal energy within the thickness of the metal. The light is absorbed by the metal, and this energizes the electrons. The electrons are accelerated by an energy field within the crystal lattice of the metal, and this generates heat. At the point at which the light strikes, the heat is so high that the material melts or vaporizes. The amount of heat generated is based on the light absorption of the material being cut. Once it pierces the metal, the cutting action begins. The beam of light moves along the contour, melting the metal, while a jet of gas blows the melted metal downward, leaving a kerf cut not much wider than the beam itself.

The surface properties of the metal influence the optical behavior of the laser beam. The peaks and valleys of a diffuse surface can trap some of the light and speed up the process. Mirror surfaces can be difficult for gas (CO_2) lasers because of the way these lasers deliver the light beam. They use mirrors and lenses to deliver the energy to the metal surface. Reflective surfaces will reflect back some of this energy into the expensive head of the laser, damaging the lens or mirrors. The surface being cut can be coated with a nonreflective material, but this can affect the final cut appearance. Fiber lasers are solid state, Nd:YAG (neodymium-doped yttrium aluminum garnet) lasers that deliver the energy by fiber. This is a very efficient method of cutting stainless steel, regardless of the finish polish.

Higher-power lasers are needed to cut thicker metal. A 2000 W laser can cut stainless steel 4 mm thick adequately. A 4000 W laser can cut stainless steel as thick as 15 mm.

One precaution with lasers is transfer corrosion from the steel slats. Steel slats are often used to support the stainless steel sheet or plate on the laser bed. As the laser pierces the stainless steel it will hit the slats that support the stainless steel sheet or plate. This in turn burns and melts the slats and can deposit steel splatter on the reverse side. This contaminates the surface and requires removal, or the stainless steel will corrode. One solution is to use copper slats or copper covers on the steel slats. The copper has a different absorption rate than stainless steel. This is important because the copper will not absorb as much energy from the laser and will not damage the stainless steel. High-powered lasers can cut stainless steel plates up to 25 mm (1 in.) thick without leaving a burr

FIGURE 6.3 Jan Hendrix art prototype cut using a fiber laser. The finish is Angel Hair.

or edge. There are 5-axis lasers that can cut stainless steel tubing and pipe, as well as forged shapes. These are specialized devices set on robotic arms. Stainless steel cuts more slowly than carbon steel, and the thicker the stainless steel the more slowly it will cut.

The edge produced on a laser should be clean and free of heat discoloration and burrs from the cutting process. Heat tint is usually not an issue with laser cutting when using a nitrogen atmosphere. If the edges do show discoloration, then it must be removed using pickling treatments.

Lasers are CNC-controlled cutting devices. They are very efficient cutting tools and can be programmed to cut very intricate shapes into both two-dimensional surfaces and three-dimensional parts. If there are a lot of piercings in a given design, some shaping may occur in the surface as heat is absorbed by the metal (Figure 6.3).

Plasma

Plasma cutting systems cut stainless steels similarly to carbon steels. Plasma cutting is typically performed on two-dimensional CNC-controlled tables. The stainless steel plate or sheet is set onto a steel lattice, and the high-energy plasma beam cuts through the metal. Plasma cutting involves creating an electrically charged ionized gas and forcing the gas through a small orifice. An electrical current is generated from a remote power source that creates an arc between the stainless steel work piece, which is given a positive charge, and the plasma gas, which is given a negative charge.

For stainless steels, the gas used is nitrogen or an argon-hydrogen mix. Carbon dioxide and oxygen gases can be used to create the plasma jet, but these will create a darkened cut.

Temperatures in the high-velocity plasma jet can be as high as 22,200°C, as this highly charged gas melts the metal and blows it away. There is not a lot of heat transferred to the stainless steel, but if there are a number of piercings you can expect warping of thin stainless steel. The high-definition plasma cutting systems reduce the heat-affected zone (HAZ) and produce a fine cut line in thin stainless steel. The handheld and less sophisticated systems will have a larger kerf and more oxidation on the edge. The kerf, which refers to the width of the material removed in cutting, leaves an edge where the cut occurs. A plasma-cut edge is rougher than the edge a water or laser cutting leaves when the kerf is made. There can be a redeposit of molten metal along the kerf. This redeposit is rough oxide that will need to be removed. For most artistic and architectural projects, plasma cutting of stainless steels has been displaced in most applications by cutting with high-quality fiber laser systems. High-definition plasmas are fast and efficient and can be considered for many projects involving the cutting of stainless steel, in particular those where post–surface finishing will occur.

Another benefit of plasma cutting is that it can be used in the field to trim and cut stainless steel. Using a straight edge or a template, the plasma cutting of large holes or shapes in the field is a viable option.

Waterjet

Cutting stainless steel with a waterjet is another method to produce custom shapes and openings in stainless steel sheet or plate. Waterjet cutting pushes a tight stream of water at very high pressure. Accompanying the water are tiny fragments of garnet. The garnet in the water cuts through the sheet or plate. Waterjets are extremely powerful and cut very thick plates. They are slower than lasers. The edge produced has a slight frosted appearance. One issue with stainless steel, particularly the mirror surfaces, is that when the jet first starts to pierce the metal, the water goes laterally across the sheet. This can make for a frosted appearance if not controlled. Cutting face down will eliminate this issue, but that solution may require protecting the surface from steel support grates that will be in contact with the stainless steel. The steel support grates may be slightly corroded, and they can transfer minute rust particles to the surface if the stainless steel is dragged over the grates.

Punching

Punch presses are a common form of cutting and perforating stainless steels. They possess the power to pierce the metal. Die clearances need to be established for the particular thickness of the stainless steel: clearance adjustments are in the 5–10% range, as compared to steels of similar thicknesses. Dies should be clean and polished and designed to pass clear through the sheet. The diameter of the holes being pierced through stainless steel should be two times the thickness of the metal at

TABLE 6.1 Cutting operations and limitations.

Cutting operation	Complexity	Limitations
Shearing	Simplest process	Straight lines; sheet and plate
Saw cutting	Straight edges and curves	Rough edge; all forms
Laser	Fast; CNC shapes	Sheet; thin plate; tubing (5 axis)
Plasma	Fast; CNC shapes; manual	Rougher edge; sheet; plate; in situ
Waterjet	Slow; CNC shapes	Frosted edge; sheet; thick plate
Punch	Fast; holes; CNC shapes	Sheet

a minimum. For instance, if you plan to punch stainless steel 2 mm thick, the minimum hole size should be 4 mm in diameter. If the holes are not round holes, then use this relationship to extrapolate the minimum dimension to be punched.

When punching stainless steel, it is important to take into consideration the work hardening nature of the metal. As stainless steel is pieced, the edge around the hole is work hardened to a greater degree than the surrounding metal. This imparts differential stresses in the sheet that will require some stress relief, or the sheet will warp excessively. With thick stainless steel sheets, this hardening can be so significant that it will pull the sheet out of the holding clamps. Die clearance and planning for internal stress buildup in the sheet material is critical for successful punching processes.

FORMING

Stainless steel can be fabricated with conventional forming and shaping equipment. There are several significant differences, though, between fabricating with stainless steels and other metals. Stainless steel has better strength than cold rolled steels and it work hardens rapidly. Austenitic stainless steels will work harden as they undergo cold forming operations. Stainless steel forming will require more energy to overcome the stiffness and work hardening behavior of the metal. More tonnage is needed to form stainless steel as compared to other metals of similar thickness.

For most simple bending operations, this matters little. But for stamping, spinning, and severe shaping, a planned approach that may require interstitial annealing steps to reduce the internal stress is needed. Interstage annealing involves heating the metal up to temperatures above 500°C to remove residual stresses that have accumulated in the forming operations; this softens the metal and allows for further cold working without splitting the metal or damaging the dies.

Roll Forming

Roll forming is a common method of producing large panels of sheet metal from coils. The method involves a series of progressive rolls that gradually shape the metal ribbon into a linear form. Each successive matching set of dies alters the shape just slightly, which causes plastic deformation along the length of the metal surface. Stainless steel requires specialized equipment to achieve successfully formed panels using roll forming systems. Stainless steel coil material entering the roll forming stations should be stress relieved to the greatest extent possible. It should be leveled in a tension-leveling set of rollers that feed into the roll forming station.

The calibration of the spacing of rolls and the matching dies must be set for stainless steel. The dies must be made to account for the higher yield and the elastic behavior stainless steel sheet will exhibit. As the metal enters the first set of forming dies, there is a stretching or slight lengthening that occurs. As the metal moves to the next station further stretching happens, and for stainless steel this is accompanied by cold working and hardening.

If the roll forming dies are not calibrated for stainless steel, they can permanently fix distortion into the stainless steel panel being formed. The best roll forming operations have multiple stations spread out over several meters to reduce the stretching that has to occur and to spread it out over the length of the station (Figure 6.4).

FIGURE 6.4 Roll forming equipment.

Press Brake Forming

Brake forming is a common sheet-metal cold forming process in which a sheet or plate is inserted into a large forming machine called a press brake. A machined and hardened die is firmly fixed into an adjustable holder. The tool or punch is mounted to the ram. The ram is powered by massive hydraulic cylinders that bring the punch down onto the sheet or plate resting on the die. The sheet or plate is formed to a given angle as the pressure is applied through the punch and through the die points (Figure 6.5).

Press brakes form material in a linear, straight line across a sheet or part. The bend can be 180° all the way back on itself, or any angle in between for the austenitic stainless steels. These are more ductile, particularly in the annealed state. Still, though, more power and force are needed to shape than with standard cold rolled steel sheets. Austenitic stainless steel can be heated to around 65°C (150°F) to allow forming at lower pressures. This temperature does not develop carbides, nor does it have an appreciable effect on the temper.

FIGURE 6.5 Custom, high-precision press brake form shapes.

FIGURE 6.6 Press brake forming using a custom shaping die.

A press brake can be fitted with special radius dies to induce straight, ruled curves into a sheet or plate. This allows for bullnose shapes and custom contours to be accurately produced in a press brake with the proper tonnage. In expert hands, a press brake is a highly versatile piece of forming equipment (Figure 6.6).

Springback

Stainless steel press brake forming requires understanding a characteristic of stainless steel known as "springback." Springback is the term used to describe the condition of a material that has undergone incomplete plastic deformation. It is a change in strain, produced by what is known as "elastic recovery." The higher the yield stress of a material, the greater the elastic recovery and springback. Springback is a dimensional change that occurs on formed parts when the pressure of the forming tool is released. Springback is a function of the material strength.

All stainless steel being formed experiences some level of springback. Springback is dependent on the ratio of bend radius and material thickness. There can also be differences in strength within the sheet itself, creating fluctuations in springback. Alloys of stainless steel with high tensile strength, such as the duplex alloys, have greater springback tendencies. It takes more power

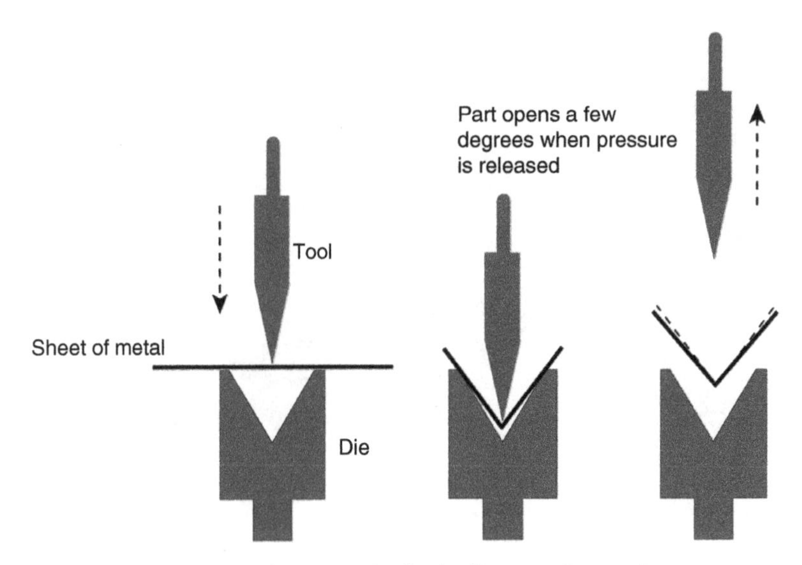

FIGURE 6.7 Overbend is necessary to arrive at a springback allowance in metal.

and force to achieve the same brake form on duplex as it does on austenitic stainless steel, and an allowance compensating for elastic recovery must be made (Figure 6.7).

Brake forming a 90° leg in a thin plate may require overbraking a few degrees to arrive at a true 90° due to stainless steel's strength and resilience. In practice, determining and gaining an understanding of minor material differences is a trial-and-error process. There are formulas that assist in approximation, but the typical practice involves first choosing a smaller bend radius and then advancing from there to a greater radius. The formed angle radius will be slightly smaller than the actual radius achieved when the pressure is released. One can also bottom out, or coin, the die, but this can cause wearing of the die material. This requires more tonnage and is not commonly used.

Air forming, coupled with modern compensation equipment and sensors, can overcome springback in material. Air forming uses the two edges of the die with the single edge of the tool. Depending on the material thickness or inside radius, the springback will change. This can be adjusted by changing the die used (Table 6.2).

Air forming overcomes springback in the sheet metal brake form by pressing the metal down into the opening of the die. Less press tonnage is required to air form. Dies used in press brake

TABLE 6.2 Bend radius and overbend requirements.

S30400 austenitic stainless steel with a 90° bend			
Bend radius	**1 thickness**	**6 thicknesses**	**20 thicknesses**
Overbend in degrees	2°	4°	15°

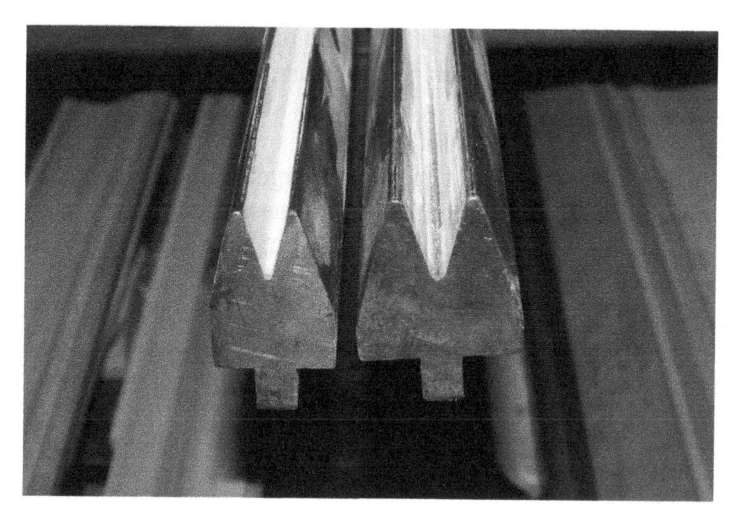

FIGURE 6.8 Air forming dies.

forming are called v-dies. They come in various angles to compensate for springback. The angle is designed to compensate for springback by pushing the metal toward the downward-moving punch. V-dies are available in widths measured from tip of die to tip of die (Figure 6.8).

As material thicknesses increase, or as the angle required increases, a die with a smaller angle may be needed to push the material toward the punch. The tool and the die must be matched to give the needed clearance. The die angle must be greater than or equal to the angle on the tool (as shown in the following table) or you may damage the tool.

90° die	Die opening range of 4–12 mm
88° die	Die opening range of 12–25 mm
85° die	Die opening range of 25–37 mm
78° die	Die opening range of 37–50 mm
73° die	Die opening range of 50–63 mm

During press brake forming the metal toward the outer side of the bend goes into tension, while on the inside of the bend the metal is in compression. Metal along the edge work hardens more than metal away from the edge. To hold its shape the metal must undergo plastic deformation, which means it will not return to its original shape with the loading is released. The metal strain hardens along the bend.

Because of the reflective nature of stainless steel, minor flaws in the dies and punches will translate to the surface of the stainless steel. If punches and dies are made of multiple sections, it is critical that these are very accurately aligned. The dies must be made of machined and hardened steel or else they will scar and dent. These defects will be transferred to the stainless steel sheet or plate being formed.

When forming with stainless steel, plastic coatings or separators are often placed between the stainless steel and steel die to prevent damage to the stainless steel and allow the metal to move as it forms. It is also critical to insure that steel particles are not transferred to stainless steel parts, particularly if the forming equipment is used for steel as well as stainless steel. Bending or brake forming stainless steel will push the bend to a slightly larger radius than other materials due to the high strength and work hardening that occurs with stainless steel. The bend radius – the radius of the curvature of the inside or concave side of the bend – of stainless steels and other high-strength materials is larger than that of softer, lower-strength metals. When the metal is formed, the other fibers (those on the outer side of the bend) are stretched, and if too tight, a form could crack. Stainless is more resilient, but for interference coated stainless steels and PVD coatings, the stretching that occurs at the outside radius can cause a slight change in the color as the coating thins over the bend.

V-CUTTING

Stainless steel can be v-cut. "V-cutting" is a term used to describe the physical removal of some of the metal at the point of a bend in order to allow for a tighter radius bend. V-cutting is a form of machining. The metal that is removed leaves a v-shaped cavity on the sheet of material before it has been formed. Specially designed equipment is used to remove the sliver of metal. A special cutting tool and tool holder are needed, as well as a vacuum table or sufficient clamps to hold the stainless steel sheet or plate firmly in position as the tool removes a thin sliver of metal. The force can be significant as the sliver of stainless steel is carved from the back surface of the stainless steel sheet (Figure 6.9).

Stainless steel is an excellent material to v-cut because of its non-galling characteristic.

FIGURE 6.9 V-cut bend (*left*) compared to standard bend (*right*).

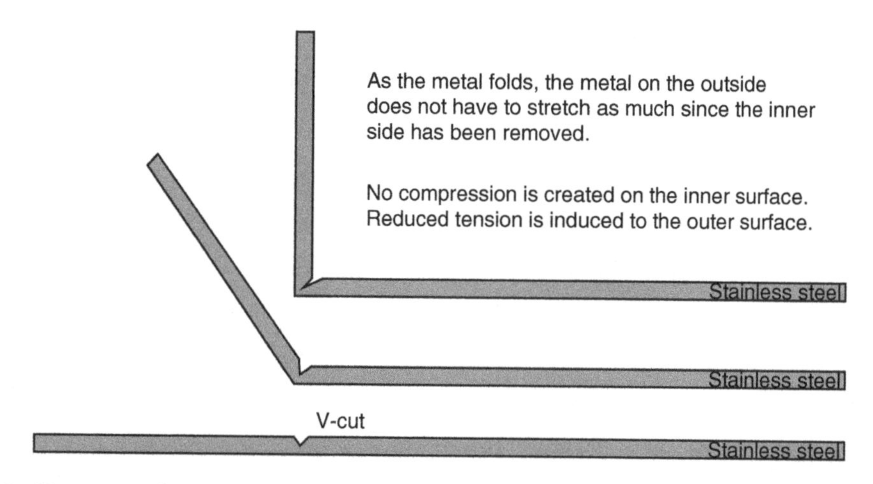

FIGURE 6.10 V-cut operation.

Properly designed tools will not add excessive heat into the sheet, eliminating the need for coolants as the curl of stainless steel is removed from the reverse side. V-cut stainless steel, when folded, produces a clean, smooth edge. If the tool is loose, chatter will be apparent on the fold.

One interesting behavior of v-cutting is the alteration of stress induced at the fold. As stated earlier, brake forming places the outer surface of the bend in tension while the inner surface is put into compression. When the edge to be formed is v-cut then brake formed, there is a significant reduction in the tension-induced stress. Compression is eliminated by the removal of the material so that there is no added contribution to unbalanced stress that might exist in the sheet as it is folded 90°. The removal of metal by v-cutting limits the compressive stress and allows for a much sharper bend (Figure 6.10).

V-cutting does weaken the corner bend. Removal of material as well as heat reduces the strength of the metal at the edge of bend. If the v-cut is too deep, or if the metal sheet is not held flat to the cutting table and a slight wave is cut into the back surface, this could lead to premature cracking.

HOT FORMING

Hot forming is usually performed on plate, bar, pipe, or tube. Hot forming stainless steel requires close control of the heat. Austenitic stainless steels will be strengthened only by cold working processes and not by heat processes (Figure 6.12).

Stainless steels are readily shaped by hot forming operations such as rolling, extruding, and forging. Curved shapes in heavy plate require hot forming processes to enable the shaping. Cold forming work hardens the metal quickly and further forming can be difficult. Heating the stainless steel form allows shaping to occur.

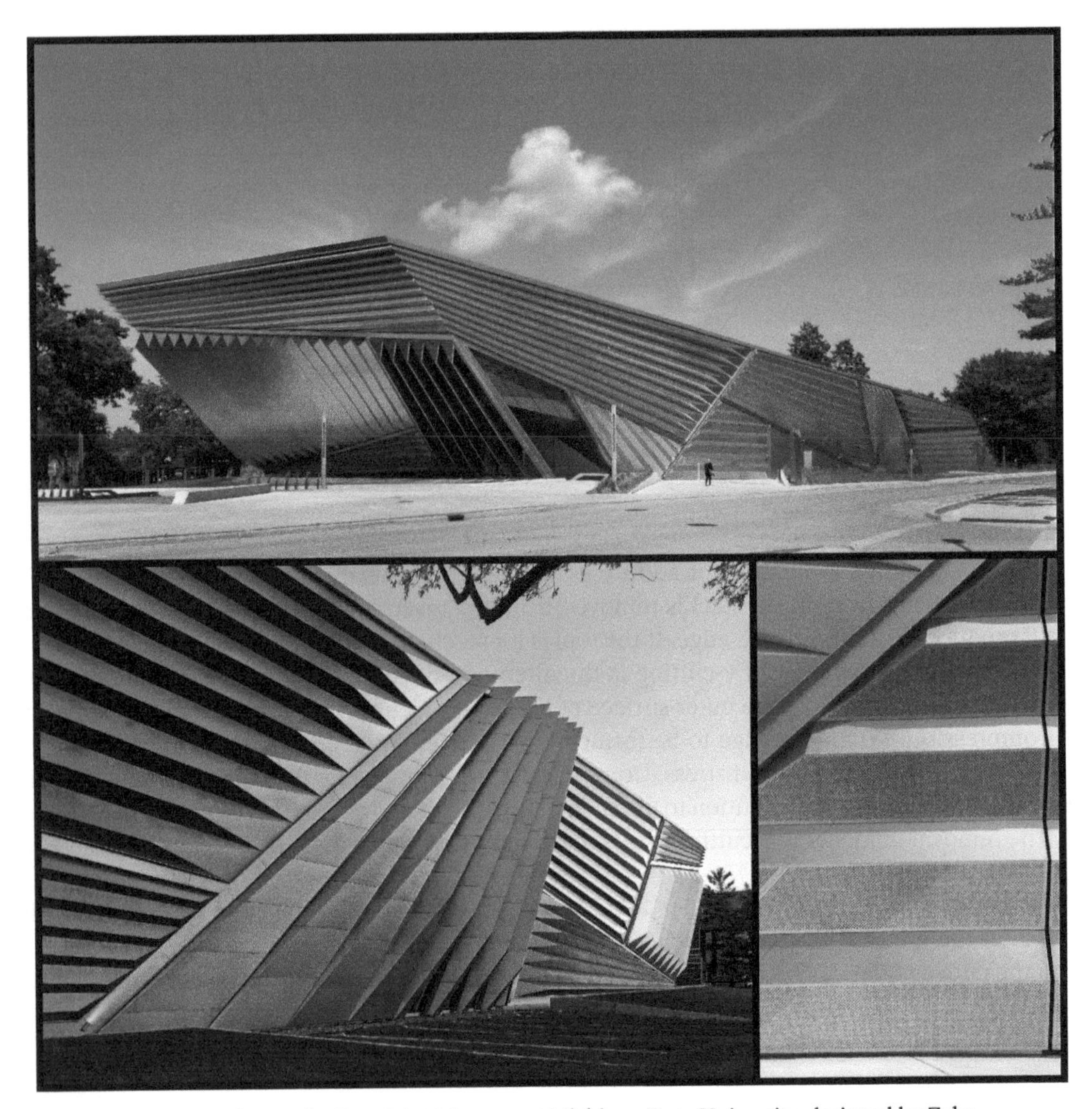

FIGURE 6.11 V-cut edges at the Broad Art Museum at Michigan State University, designed by Zaha Hadid Architects.

FORGING

Forging stainless steel involves heating the metal and applying force rapidly to the surface. The challenges with stainless steel are several. Stainless steel work hardens rapidly, making it more difficult to form. Each subsequent impact to stainless steel requires several times more energy than for carbon steels.

FIGURE 6.12 Hot formed curved and tapered stainless steel pipe, designed by Vicki Scuri.

In art and architecture, S30400 or S31600 are normally considered for forging. These stainless steel alloys can be annealed. To forge these alloys, the temperature is held within a range of 926–1260 °C. To soften the metal during working, anneal temperatures are kept above 982°C, then rapidly lowered to 480°C for quenching.

In forging stainless steel, there is a limit to the number of successful annealing operations. After three or four the metal's properties will begin to alter.

MACHINING

Stainless steel alloys can be machined with the right equipment, tooling design, and speed rate. It is more difficult than for steels and nonferrous materials because of its higher strength and how rapidly it will work harden. The austenitic alloys are machinable, as are the ferritic alloys. Martensitic alloys are more difficult, due to their inherent abrasive properties. The tools wear quickly with martensitic and precipitation-hardened alloys. The equipment used to machine stainless steel must be robust

FIGURE 6.13 Tooling used for milling.

and rigid. Chatter and vibration should be eliminated. The tools used to cut the stainless steel must be sharp and designed to create small chips rather than long spirals (Figure 6.13).

Austenitic stainless steel has a tendency to mill off in large spiral coils that can wind around a tool. The chips will take the heat away. Coolant can be used to assist in moving the chips away, or an air blast will work as well. The tools should be tungsten carbide–tipped or coated with titanium aluminum nitride. High-speed steel with cemented carbide will also work well.

When machining stainless steel, the cuts should be deep and made at lower speeds than those used to machine steel to prevent work hardening from developing. There are free machining alloys of stainless steel that allow for increased rates of cutting. When machining stainless steel, avoid glazing over a surface; this will work harden the surface and make it difficult to machine.

The stainless steel alloys best suited for machining have a higher allowable sulfur content. Alloys such as S30300 are designed for machining. The sulfur is present as manganese sulfide nonmetallic inclusions. The sulfur aids in tool life and allows for faster speeds. The finish clarity and corrosion resistance are reduced with the sulfur intermetallic. The selenium-bearing alloy S30323 provides better ductility than an alloy with sulfur alone, and therefore better machining, with the added benefit of improved finishing characteristics (Figure 6.14).

SOLDERING AND BRAZING

Soldering and brazing are similar processes that use an additive metal to join two metals together. In both processes, molten metal is applied to the joint and allowed to solidify, and the metals are joined at the joint. This creates a metallurgical bond between the additive metal, the filler, and the surfaces of the metals being joined. This metallurgical bond is only as strong as the filler metal used.

FIGURE 6.14 Examples of various custom-machined stainless steel parts.

Unlike welding, soldering and brazing processes are lower-temperature processes that can be reversed. You can add heat energy, melt the joint, and separate the two metals. The distinction between soldering and brazing lies in the temperature used to melt the filler metal and join the parts, as expressed by the following formula:

$$\text{Soldering} < 450^\circ\text{C}\ (840^\circ\text{F}) > \text{Brazing}$$

The success of soldering and brazing processes depends on the “wetting” ability of the filler metal. Wetting refers to the ability of the filler metal to be drawn into the joint sufficiently to make a good, metallurgic bond. Fluxes are used to facilitate wetting behavior. A flux is a compound that removes the oxide on the base metal. Typically, a flux is a chemical compound, often acidic, that when applied degrades the oxide by weakening the bond to the base metal. The flux penetrates the oxide though natural porosity of the oxide layer or tiny fissures or imperfections in the oxide layer. The flux remains on the surface during the process of joining and retards the development of a new oxide layer. It can also move the heat and draw the molten metal by reducing surface tension. This allows the molten metal to join to the base metal and displace the flux.

Fluxes by their nature are corrosive, and for stainless steel it is very important to remove any excess flux from the joint after it has been soldered and brazed. Residual flux will lead to corrosion at the joint.

Both processes involve heating the base metal and the filler metal. The process introduces lower heat than welding processes; however, what often occurs is that the molten metal expands the region around the joint, then when this solidifies, the distortion is held and permanently fixed into the part. As the molten metal cools, it pulls the thin stainless steel surface toward it and creates a convex indentation on the reverse side. It is very difficult to arrive at good, distortion-free joints when soldering or brazing.

Thin stainless steel sheets can be seamed together and soldered. Flux is required to remove the chromium oxide layer from the surface of the area being joined. Flux allows the melted metal solder, usually a tin-lead alloy, to wet the joint and flow into the seam, bonding the metal surfaces. The use of special soldering fluxes made for stainless steel is recommended. Zinc chloride fluxes are still commonly used to remove the oxide from stainless steel, but it is critical to removal all excess flux from the solder joint. For overlapping seams this can be difficult to impossible. Excess flux will corrode the stainless steel in a matter of a few days. So, flushing the joint and removing all flux is critical. There is phosphoric acid–based fluxes that are less harmful to stainless steel, but they act more slowly to flux the joint. Solder used on stainless steel can be any of the lead-tin or tin-antimony solders.

Stainless steel has poor thermal conductance properties as compared to other metals, such as copper or tin. When heated to soldering temperatures, usually near 205°C, the heat is concentrated and will build more quickly than soldering copper sheet. The drawback is more in appearance. The metal thermally expands in the area around the solder, rendering the surface wavy.

Brazing is a higher-temperature process used to join two metals together. Although more commonly used with copper alloys, brazing of stainless steels is possible, but controlling the heat-induced warpage is challenging when brazing flat surfaces. Tubes and pipes are more effectively brazed due to their inherent section.

There are three categories of brazing used on stainless steel: air with a flux, brazing in a reducing atmosphere, and brazing in a vacuum. For all categories, the area being joined must be thoroughly cleaned. Brazing temperatures are in the range of 600–700°C. Metals used as brazing fillers are alloys of silver, nickel, and copper. These have higher melting points than tin-lead solders. Fluxes are needed to achieve flow into the braze joint, but an inert atmosphere is also needed to restrict the development of chromium oxide. There are industrial ways of brazing stainless steels in controlled atmospheres and vacuums, and there are high-temperature fluxes available that dissolve the oxide. There are a number of fluxes used for brazing stainless steels. They are solid, paste-like, alkali metal salts that melt just below the melting point of the brazing filler. As the joint is filled with the brazing metal, the flux is pushed out. Once the joint is brazed together it is thoroughly cleaned of all fluxes. Post-finishing would follow.

WELDING

Welding of stainless steel is in common practice for all wrought and cast types of stainless steel and all alloys. The austenitic alloys are the more accommodating of the stainless steel alloys to weld. All of the welding processes can be used on the austenitic stainless steels. There are particular behaviors unique to welding stainless steel when compared to their cousins the carbon steels.

Unique Challenges to Stainless Steel Welding

- Carbide precipitation
- Lower melting point than steel

- Greater electrical resistance than steel
- Lower coefficient of thermal conductivity than steel
- Greater coefficient of thermal expansion

The HAZ is the area adjacent to the weld that receives a significant amount of heat during the welding process. In this region, carbides can precipitate out of the steel alloy, combine readily with chromium, and create chromium carbides that will damage the stainless steel's corrosion resistance. It is not the chromium carbides themselves that do the damage, but the chromium depletion that occurs on the surface when this region reaches a high temperature of around 650°C. At this temperature, chromium carbides form quickly. These weakened areas must be removed in a passivation/pickling process. Today, most alloys have a reduced carbon content. The low carbon alloys S30403 and S31603 have a maximum carbon content of 0.03%. Using low carbon alloys will help to reduce chromium carbide development.

Stainless steel has a lower melting point than steels. Carbon steel melts at around 1540°C, whereas stainless steel has a melting temperature of around 1425°C. This allows for a small reduction in temperature to fuse the metal.

The electrical resistance of stainless steels is greater than that of carbon steels. This means lower electrical current is needed to generate the welding heat. Carbon steel has a resistance of 12 μΩ at room temperature, compared to stainless steel with a resistance of 72 μΩ at room temperature. When the temperature of the metal has increased to around 890°C, carbon steel and stainless steel will have about the same electrical resistance of 125 μΩ.

Stainless steel has a thermal conductivity significantly lower than that of carbon steel. Heat is not dissipated away from the weld to the same degree as it is in carbon steels. This allows welding current to be reduced, but warping and elongation can result, making it difficult to join edges. A process of skip welding and chill bars is often introduced to aid in stainless steel welding practices by reduce heat concentrations and by heat dissipation.

Stainless steels have a coefficient of thermal expansion 50% greater than that of steels. This can increase warping around the HAZs. Using techniques to remove the heat artificially with chill bars will reduce the warping and distortion that can accompany the heat of welding.

The goals in welding stainless steel are:

- Achieve a strong connection
- Reduce warping
- Sustain the corrosion resistant behavior
- Resist cracking from stress as the metal cools

There are several processes for welding stainless steel, and each has a set of criteria that go beyond what can be covered here. Perhaps one of the only processes that is used on steel and not on stainless steel is oxyacetylene welding. Oxyacetylene welding would add carbon into the weld zone, and this would render the area around the weld prone to corrosion. Welding processes can either

FIGURE 6.15 *Winged Refuge*, an Air Force memorial sculpture by artist John Labja, being welded and polished.

add metal to the joint or fuse the joint together without adding metal. When metal is added in the welding process, it is called a filler metal. It is critical that the alloy of this filler metal is very close to that of the base metal in composition. In welding duplex alloys, the filler metal should be higher in nickel to maintain the close ratio of austenitic to ferritic makeup.

Process improvements, use of robotics, and new heating techniques such as induction are changing the way stainless steel is welded. The point of welding is to join one surface to another and to arrive at a system with good strength and corrosion resistant properties equal to that of overall body. One kind of welding is resistance welding, which is a forge welding process in which pressure is applied as the heat of melting is achieved. This is also called spot welding. There are also several other processes used that are fusion welding processes, where the metal is melted and fused, or metal is added to fuse two sections together. These are discussed in the sections that follow.

Fusion Welding Processes Used on Stainless Steels

Table 6.3 lists the different fusion welding processes used for stainless steel and compares the advantages and disadvantages of each. These processes are described in detail in the sections that follow.

TABLE 6.3 Comparative weld processes used on stainless steel.

Welding process	Advantages/disadvantages
Gas tungsten arc welding	Low heat, thin metals; automated and manual; requires very accurate fixturing; single-side
Gas metal arc welding	Fast and efficient; automated and manual; no flux needed; low oxide and little slag buildup; thicker sheet; more weld cleanup than gas tungsten arc welding
Shielded metal arc welding	Rapid and versatile process; automation of process is difficult; slag and flux cleanup required
Flux core wire welding	No shielding gas needed; can be automated; cleaning difficult; higher temperature.
Submerged arc welding	No shielding gas needed; high-productivity; high heat buildup; creates more ferrite in weld; thicker plate only
Plasma arc welding	Good weld quality; filler metal is added; automated system; thin and thick stainless steel; generates high heat
Electron beam welding	Excellent weld quality; expensive system; in-factory process; very accurate and low heat input; both thin and thick stainless steel can be welded; requires a vacuum
Laser beam welding	Excellent weld quality; expensive system; in-factory process; automated; can weld thick or thin sections depending on power output; minor cleanup
Capacitor discharge welding	Thin plate stud welding; show-through to face side challenging
Arc stud welding	Heavy plate stud welding; not suitable for thin sheet
Friction-stir welding	Fuses plates from heat of friction; limited use

Gas Tungsten Arc Welding

Gas tungsten arc welding is also known by the initialism GTAW and as TIG (tungsten inert gas)welding. In this process a high-energy arc is created between a tungsten electrode, which is not consumed in the welding process, and the metal. This process can produce a weld that requires very little cleanup. A major benefit to this welding process is lower heat input into the stainless steel joint, which leads to lower oxide and slag levels. GTAW can be performed in all angles and all positions. It is a single-side welding operation.

Metal wire made of the appropriate alloy can be fed into the arc at the joint to be welded. The wire is fed though a torch, which also distributes a shield of inert gas around the area of weld. Metal can be added manually by feeding it into weld zone. Metal can also be fed into the weld zone automatically by using a timed feed system. GTAW can be manual, semiautomated, or fully automated (using robotic welding systems). GTAW is suitable for welding thin stainless steels between 0.3 and 3 mm in thickness (Figure 6.16).

Gas Metal Arc Welding

Gas metal arc welding is also known by the initialism GMAW and as MIG (metal inert gas) welding. It is similar in most ways to GTAW, with the exception that the wire is continuously fed and creates the arc without tungsten. The wire electrode is consumed as the welding proceeds. The filler metal is automatically fed into the weld zone. GMAW is a very fast and efficient process that can be semiautomated or automated. Systems can be set to weld both thin and thick stainless steel. The most optimum stainless steel thicknesses are between 2 and 10 mm. A shielding gas is used to keep the weld clean and free from oxygen and nitrogen contamination from the surrounding air. No flux is needed. The process produces very low oxide and slag. GMAW can be undertaken in all positions and on both sides of a part. This system is common for welding stainless steels.

Shielded Metal Arc Welding

Shielded metal arc welding is also known by the initialism SMAW and also as stick welding. In this process a high-energy arc is created (struck) between a current-carrying electrode and the metal to be fused. The electrode feeds stainless steel wire to the joint to be welded. The wire has a coating of flux that melts in the high temperature of the energy arc and provides protection and cleans the area directly at the weld. The high heat energy melts the metal being fed into the joint and the metal being welded. This is a common manual approach. It is not easily set for automation.

Flux Core Wire Welding

Flux core wire welding (FCW) uses a hollow wire containing a high-temperature flux. The wire is passed through a welding apparatus and enters the high-energy arc. No external inert gas is

FIGURE 6.16 Building at 200 11th Avenue, designed by Selldorf Architects.

required because the flux reacts to develop a shielding envelope around the weld. This method is beneficial when it is necessary to weld in windy conditions and the shield gas can be blown away, because FCW produces the shield gas on the target weld as the weld is laid down. Cleaning the weld is more challenging, however, due to the introduction of the flux at the point of the weld.

Submerged Arc Welding

In submerged arc welding (SAW), a stainless steel alloy wire electrode carries a current and creates the high-energy arc. Flux is added by means of a separate feed into the joint being welded. The arc is created within the flux cover. This is a high-productivity, automated process used on mainly flat surfaces. Heat input is high when using this process, so it is more suited to heavier thicknesses (10 mm or thicker). When welding austenitic stainless steels, care must be exercised to reduce ferritic development within the weld.

Plasma Arc Welding

Plasma arc welding (PAW) uses a special welding torch that constricts the high-energy arc by means of a special nozzle. This increases the temperature and concentrates the heat. Like the GTAW process, bare metal wire is added and consumed at the weld joint. The plasma jet is very high temperature, around 28 000°C (50 432°F). Depending on the weld type, PAW can weld thin stainless steel of less than 2 mm and thicker plate up to 25 mm.

Electron Beam Welding

Electron beam welding (EBW) is a specialized process performed in a vacuum. The EBW process aims a dense stream of electrons at the area to be welded. The high energy of the electrons is converted to heat sufficient enough to melt the metal. Usually the metal joint is designed so that this very fine stream of particles fuses the metal edges together without the addition of filler metal. Thicker metal can be welded to thinner metal. The EBW system produces low distortion in the welded parts. You can very accurately control the depth of the weld penetration and the heat is minimized. This produces very low distortion. This factory-automated process requires a high vacuum, which eliminates oxidation. This is not a common welding process for art and architecture.

Laser Beam Welding

The process of welding stainless steel with a laser focuses a fine, high-energy beam at the point to be fused. The metal absorbs the energy at that exact point and melts from the tremendous heat that is developed. The main benefit is the reduced HAZ. The laser can be set to provide just enough energy to melt the metal. This reduces distortion and confines the area affected by the weld to a minimal zone. Properly welded joints with laser will require very little cleanup. The challenge is fixturing the parts to be welded. Laser welding requires CNC control or robotic control. The accuracy of precise fit-up and laser path layout cannot be ignored.

For laser welding, low carbon alloys and low sulfur alloys are suggested. Due to the small area of heat, the metal cools rapidly. Some of the elements in the alloy will come out of solution quicker and solidify, causing embrittlement around the weld.

Laser beam welding is a factory process that can be used on stainless steels of varying thickness depending on the power. Usually CO_2 lasers are used for welding due to their higher power, although

Nd:YAG lasers can weld thinner sheets just as well. Powerful CO_2 lasers can weld plates up to 38 mm thick and are used to produce stainless steel structural forms.

Capacitor Discharge Fusion Stud Welding

This is a method of attaching a stud to a stainless steel surface by means of a high-energy discharge. The capacitor discharge fuses small-diameter, threaded and unthreaded studs to thin sheet stainless steel. The stud is held against the surface as a capacitor is activated to release stored energy. The energy is sufficient to melt the face of the stud and the surface of the sheet material. The stud is pushed into the molten metal, which cools and fuses the two surfaces. It happens very rapidly, and distortion is reduced.

Arc Stud Welding

In the arc stud welding process, heavier stainless steel plate is used as well as larger studs. A ceramic cup surrounds the base of the stud at point of contact to the plate. The stud is raised just off the surface as energy is passed between the parts. The ceramic cup concentrates the heat in the area just around the weld and reduces weld splatter. When the base metal plate surface is molten and the surface of the end of the stud is molten, a spring-loaded ram shoves the stud into the molten metal. Solidification occurs rapidly, and the stud is set perpendicular to the surface.

Friction-Stir Welding

The friction-stir welding process was developed in Cambridge, England, in the early 1990s. The process involves a high-speed tungsten alloy tool that spins as it moves between the two edges to be welded. Used extensively on aluminum, friction-stir welding has shown success in welding plates of stainless steel together as well. The spinning tungsten tool generates heat from friction sufficient to melt the stainless steel. As the tool passes between the two edges, it melts the metal immediately around the spinning tungsten tool and the trailing metal resolidifies as it cools, effectively joining the plates together. This welding technique is not in regular use by metal fabrication companies involved with art and architecture.

The Steps in Welding

For successful welding of stainless steel, there are a series of distinct steps that when planned and followed will deliver the results intended. The steps and processes vary depending on the alloy, so one who is versed in the welding of austenitic alloys will have difficulty welding duplex alloys if adjustments in process are not undertaken.

All the processes used to weld stainless steel require planning in order to be successful. The area where welding is to take place must be clean and free of contamination from other metals, in particular steels. Tools used to grind and prepare the stainless steel for welding must be dedicated to stainless only. Fixturing and support for holding the material to be welded must be rigid and clean. Chill bars, protective atmospheres, and proper weld techniques must be used to aid in cooling and achieving a finish weld that meets project requirements.

Joint Preparation

Beveling joints in thick stainless steel allow for more passes with less heat buildup to create the joint. It is critical to insure that oils and grease are not on the edge. These can contaminate the weld and affect the corrosion resistance by increasing carbon. Waterjet-cut edges need to be treated before welding to clean the garnet fragments that can be left embedded in the edge. Keep all carbon steel brushes and files away from the stainless steel to avoid contaminating the surface.

Filler Metal

Filler metal is the added metal introduced into the weld. Using the correct filler metal is important. When welding austenitic and ferritic alloys, you want the filler metal to be an alloy similar to or matching the base metal.

Welding Duplex Alloys

Duplex alloys require filler metals that have a higher nickel content and a lower nitrogen content than the matching alloy. The duplex alloy S32205 has appropriate minimums of nitrogen along with increased minimum allowances for chromium and molybdenum. Combining it with a filler of higher nickel content will help to develop the right level of austenitic and ferritic mix at the welds. The filler metals used on duplex alloys will have an increased amount of nickel. For S32205 one should use a S32209 alloy filler rod, which has 2% more nickel than the S32205 alloy. It also has high percentages of chromium and molybdenum, which further increase the corrosion resistance of the weld. The increased nickel content is needed to keep the duplex phases of austenite and ferrite close to the composition of the base metal after the weld cools.

Heat Input During Welding

On thick stainless steel, multiple passes are necessary to lay down the correct amount of weld. There are specific heat ranges that should be used for each of the various alloys.

Duplex alloys have a higher thermal conductivity coupled with a lower coefficient of expansion, whereas the austenitic alloys develop intensive stresses related to local thermal conditions at the welds, which can create cracks.

For duplex alloys, multi-pass heating can create chromium nitride precipitation around the HAZ. For duplex alloys it is important to observe temperature limits. Duplex alloys require a higher heat-input temperature and lower limits on interpass temperatures. Maintaining these limits will help to keep the balance of the duplex phases intact at the weld site.

Post-Weld Treatment and Cleanup

For the duplex alloys, heating the surface to relieve post-weld stresses is not necessary and may affect their toughness and corrosion resistance by changing the phase of these alloys. For the austenitic alloys, higher temperatures in post-weld annealing are required because of the higher nickel content.

Post-weld cleanup is critical for stainless steels. Cleaning off any oxides that have formed around the joint is highly recommended. This should be done chemically, using selective electropolishing techniques or any number of special pickling pastes and passivation gels. Use the process that is easiest and most effective, as well as the most conscientious in terms of safety and the environment. Pickling pastes and gels come in different strengths. It is important to reduce the nitrous fumes and keep workers safe when using many of these pastes.

Removal of any carbon precipitation at or around the weld will contribute to the long-term performance of the stainless steel surface. Refrain from using steel brushes of any sort. Grinding off the discoloration can appear to indicate success visually, but there may still be remaining weak areas of chromium depletion. Use of low carbon alloys will benefit post-weld cleanup by reducing the chromium-depleted zone around the weld. Still, it is advisable to utilize chemical or electrochemical pickling and passivation techniques. See Chapter 8 for more information on passivation.

Removal of weld splatter is necessary to eliminate areas where undercuts or snags could capture and hold contamination on the surface. Additionally, these weld remnants may be chromium depleted and lead to eventual corrosion.

Distortion

From an aesthetic viewpoint, weld distortion is the most common challenge faced with stainless steel. Distortion can be introduced by the heat input of the process and subsequent rapid cooling or can be imparted by grinding and weld-reduction techniques that rely solely on a human's ability to apply force and maintain the position of a rapidly moving grinder. If the elements being joined by welding are curved and arching sections, cleaning down the welds can be a significant challenge. With straight, flat sections, the cleaning down of the welds can be staged using tapes and block sanding. Not so with curved surfaces. These will require very careful removal, using gauges and curved protection barriers.

One of the most amazing welded and polished surface ever attempted in art is *Cloud Gate*, the stunning sculpture in Chicago by Anish Kapoor. The stainless steel plates were shaped and braced in special stainless steel forms cut and assembled to the shape. The thickness of the stainless steel is 10 mm. Each of the more than 160 pieces was custom shaped, shipped to the project site, and assembled. Initially they were joined by tack welds at the intersections. But the artist wanted all joints to be welded, ground, and polished. The welding technique used was a hybrid laser system. Argon was used to keep the welds clean and cool the joints as the welding occurred. This way distortion was kept to a minimum by reducing the passes by accurately setting out the joint spacing. This was just half the challenge. Grinding back the weld and polishing the curved surface was an equal challenge: had the joint been ground back too far, distortion would have appeared in the mirror surface. Making all this more complex, the process had to occur at the site and not in a controlled environment, such as a factory. The artwork was encapsulated in a tent for several months. This kept wind and dust from the outside environment from interfering with the process, and it allowed for the creation of a local environment within the tent. The grinding and polishing went through several very precise steps of carefully taking the weld down, then shaping the weld and blending the weld, followed by sanding with ever-smaller grit sizes. The final step required buffing with jeweler's rouge to remove the small grit lines induced from the sanding processes.[1]

Joint design, welding technique, preparation, grinding, and finishing techniques are important considerations to be worked out for a successful result. Prototype construction, to prove out technique, is a good place to begin and learn from when approaching complex fabrication using stainless steel. Constructing a prototype will establish the course of action necessary to achieve the desired results. Designing devices to facilitate removing the heat generated during the welding process can be one of the critical steps in achieving a distortion-free surface.

The metal surface of the building at 200 11th Avenue along the High Line in West Chelsea, New York City (designed by Anabelle Selldorf) is only 2 mm thick. These 2-mm-thick plates of S31603 stainless steel shapes were assembled from waterjet cut plates that are curved and welded along the seams. To achieve the clean-cut shapes and edges, the metal was placed on aluminum forms. The aluminum forms had grooves created at the point of weld where argon could be pumped through and flood the weld zone. A robotic TIG welder was utilized. It was programmed to follow the contours and produce a very accurate fused joint. To achieve this, the edges of the thin plates of 2-mm stainless steel were prepared to come to an accurate meeting point along the curve and tack welded very carefully by hand using a MIG setup in which matching alloy wire was deposited in small tacks.

Waterjet edges can be contaminated with the garnet used to facilitate cutting the shapes. Post-cleanup of the edge was critical: the accurate tapering of this edge was accomplished by passing the surface over a fixed grinder and creating an angled bevel. The weld achieved was accurately and consistently applied. Post-cleanup of the weld was minimal. The edge of the stainless steel form undulates across the Selldorf design like a repeating wave. Success of the design was dependent on the success of the weld and the post-cleanup of the surface (Figure 6.16).

Steps to Reduce Distortion in Welded Surfaces

- Straightened and smooth plates
- Prepared and cleaned edges to be welded
- Testing and fit-up of joints before weld application
- Adding shape to the surface before welding, if possible

[1] *Cloud Gate*, designed by Anish Kapoor, was fabricated by Performance Structures, Inc. of Oakland, California, and MTH Industries out of Chicago. In the author's opinion, next to the Gateway Arch in St. Louis, *Cloud Gate* is one of the most amazing metal sculptures ever conceived and constructed.

- Locate weld at edges
- Reduce tack weld size
- Continuous checking of fillet welds using fillet gauge
- Use of automated systems
- Incorporate pattern weld techniques to reduce heat build
- Use of heat sinks in welding process
- Step grind and use gauges such as tapes to prevent over grinding
- Use several steps in sanding starting with larger grit down to smallest grit
- Test process on prototype assemblies

CASTING

The cast form of stainless steel is becoming more commonly used in art and architecture. The processes used to cast stainless steel are similar to those used in casting steels. The foundries that cast stainless steels often are the same ones that cast steels. Cross-contamination prevention is an important quality control criterion to establish early in the development process for cast stainless steel.

Listed in Chapter 2 are several alloys commonly cast for art and architectural projects. These alloys are austenitic forms with attributes similar to wrought versions. It is important to note that the microstructure of cast stainless steel is different from the microstructure of wrought forms. The microstructure of cast stainless steel is determined by the way the molten metal cools and solidifies in the mold, whereas the microstructure of wrought forms of stainless steel is determined by the annealing and cold working processes that extend and align the microstructure.

This distinction will make matching color tone and finishing more difficult, if not impractical. The color and reflectivity are determined by the microstructure on the surface of the stainless steel casting or wrought version.

Austenitic Cast Alloys

With stainless steel, it is not always practical to cast large parts. Usually castings are less than 100 kg (220 lbs.). In practice, after casting it is important to heat treat the cast stainless steel to arrive at the desired properties. Heat treatment for austenitic alloys involves solution heat treatment at temperatures of 1040–1150°C and then rapid cooling in water. The ACI number designations for the cast austenitic alloys most commonly used in art and architecture are J92500, J92600, J92701, J92800, and J92900.

When casting these alloys, a significant amount of ferritic microstructure forms, sometimes as much as 20% of the casting. There is no easy way around this. The ferritic portion develops as the metal cools and is usually connected throughout the casting. There are benefits as well as concerns related to the inclusion of the ferritic microstructure. The ferrite inclusion improves strength and weldability, but the ductility and toughness are reduced. Corrosive attack can penetrate and move through the ferritic portion more readily than the austenitic portion.

These ferritic pools can also develop from welding of cast stainless steel, as carbides precipitate out more readily and need to be removed in subsequent passivation procedures.

Machining castings can be more difficult than machining wrought stainless steel forms. There may be variations in the casting, oxides, and mold materials that can dull the cutting tools. Cutting speeds should be lower on cast forms of stainless steels.

Common Cast Processes Used

The most commonly used processes for casting stainless steel used in art and architecture projects are sand casting and investment (lost-wax) casting. These techniques are described in the sections that follow.

Sand Casting

Sand casting is the most common technique used for casting stainless steel parts for art and architecture. Stainless steel casts in ways similar to steel. The foundry techniques, sand types, and process involved with delivering the metal and pouring the metal into the mold are identical. The one big difference is contamination prevention. If a foundry works with steel, the potential of cross-contamination when processing stainless stain is significant. Final cleanup of the stainless steel surface should be performed in an isolated area with equipment devoted solely to stainless steel.

The mold used in sand casting is made from various types of refractory sand that has been compacted around a pattern. The most common type of sand used to create the mold is referred to as "green sand." Green sand is a granular refractory sand with a mixture of clay and other additives to help bind the grains together and give the mold strength and hardness. The compactness of the mold determines the accuracy of the final form and the tolerances possible with green-sand casting processes. The pattern can be various forms developed in the shape of the final casting. Wood, foam, and (more often today) rapid prototyping[2] are methods used to create the shape to be cast. These must be able to handle the pressure of compaction as the sand is tamped down around them to create the mold (Figure 6.17).

For rapid prototype patterns, the shape is attached to a pattern board and the sand is packed around it. In sand casting a mold and core can be created by directly fabricating from a computer model. The rapid prototyping produces the mold and cavity layer by layer using a chemically activated sand.

[2]Rapid prototyping is a general term used to describe various methods of creating objects directly from a computer model. The three-dimensional computer model is converted to an .STL file. This is the format used to produce an approximation of the three-dimensional surface that defines the shape by triangles. The more complex the shape, the more triangles are needed to sharpen the resolution of the part.

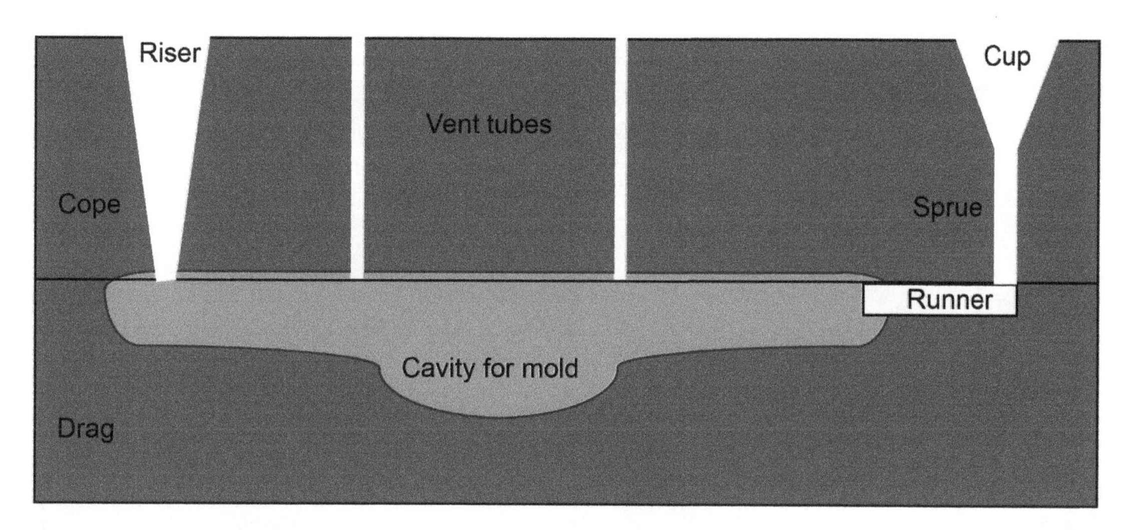

FIGURE 6.17 Parts of a sand cast mold.

Another method used is to define the shape in a three-dimensional model and carve the shape out of a block of compacted sand. The shapes are limited in complexity using this method, but they speed up the mold creation.

The Importance of Cleaning

In the sand casting process, the mold is destroyed when the cast part is removed. The surface of the stainless steel is dark and partially contaminated with refractory material. Stainless steel sand casting will require cleaning and descaling of the surface. If blasting with sand is used to remove this first layer of scale and oxide, then pickling and passivation must follow. It is necessary in order to bring the surface of the stainless steel to a passive, clean state.

The texture of the refractory sand will establish the surface quality and texture of the stainless steel casting. The most basic surface is similar to a heavy glass bead texture. The final surface can be electropolished to brighten and insure all oxides are removed. This can soften the surface as well (Figure 6.18).

Investment (Lost-Wax) Casting

Lost-wax casting, on which the modern investment casting process is based, has been used for creating intricate cast parts in metal for centuries. The process involves creating a shape from disposable wax or plastic material. Once created, the shape is turned into a mold by covering the surface with a ceramic material. Gating and venting systems are attached, and the mold is dipped into a slurry of refractory material sometimes containing silica sand or zircon. The piece is allowed to dry, then it is further coated in more of the slurry until a thick, durable shell is created. The ceramic shell is

FIGURE 6.18 Cast, hand-forged, and machined 316L stainless steel feather by Reilly Dickens-Hoffman and R+K Studios.
Source: Photo courtesy of Kris Vaswig.

fired to create a sound mold capable of receiving the molten metal, and this heat removes the wax or plastic from within the mold.

The advantage of this method is that the finish surface is smooth and requires less post-processing to arrive at a good surface. Machining of the surface is eliminated or greatly reduced. The dimensional tolerances are also superior to sand casting. The major drawback to this

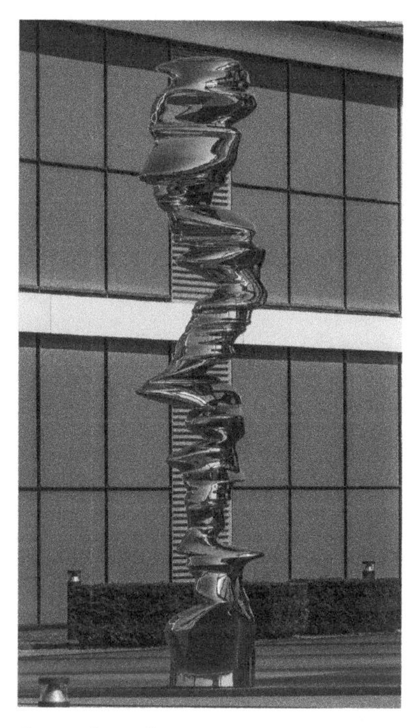

FIGURE 6.19 Cast stainless steel sculpture by artist Tony Cragg.
Source: Photo by Reinhold Möller, used under CC BY-SA 4.0, https://creativecommons.org/licenses/by-sa/4.0.

method is limitation in size. The difficulty in producing large shapes in a single mold is a constraint most cast foundries cannot overcome. The approach is to make pieces in smaller castings and join them together by welding and then post-polishing the welds (Figure 6.19).

Rapid prototyping the shape from a three-dimensional model is allowing this process to be a much more practical approach for art and architecture. The mold is produced directly from the computer by printing the form using thermal polymers or other substances that replace the wax shape and the slow creation of the wax shape.

CHAPTER 7

Corrosion Resistance Characteristics

"Corrosion": from the Latin word "corrodere," meaning "to gnaw to pieces."

An artist or designer considering metal has the choice of using a metallic paint coating over aluminum or galvanized steel. This will achieve only a temporary color, since all paints will eventually struggle against entropic forces of ultraviolet radiation, weathering, and thermal changes. The designer can choose a metal that changes with time such as copper, weathering steel, and zinc, and accept the patina color and variations of oxide that will develop with time and exposure. Or the designer can choose a metal that changes very slowly, almost imperceptibly.

Stainless steel offers a predictable long-term appearance. The major attribute of stainless steel, its metallic luster, will be expressed for a very long time in most environments. This hallmark of stainless steel is due to its passive nature. Passivity is a characteristic of all metals. At any given time metals are in either a passive state or an active state, depending upon the environment they find themselves in. Stainless steel produced correctly at the Mill source is delivered to the manufacturing operation in a passive, nonreacting state.

Stainless steels are considered engineered alloys. They were developed from steels through the addition of certain alloying constituents, particularly chromium, to achieve beneficial qualities of appearance and corrosion resistance. The element chromium is very stable. It does not want to react with many other substances other than oxygen. When alloyed with iron in amounts greater than 10.5%, chromium lends its corrosion resistance and protects the iron from chemical degradation.

Passivity for stainless steel is determined by the soundness of the chromium oxide protective layer that develops on the exposed surface of stainless steel. This thin, transparent film on the surface is enriched in chromium oxide. This intervening oxide film inhibits the free flow of ions

FIGURE 7.1 The roof surface of Independence Temple (previously called RLDS Temple) after over 25 years with only natural cleaning. Designed by HOK.

from the base metal into the environment. The amazing part is that this protective oxide is only 0.0000025 mm thick. Compare this to the thickness of a human hair, which is 28 000 times thicker.

Where stainless steel runs into challenges is in high-chloride environments: that is, humid coastal regions or areas near northern urban streets that apply abundant deicing salts to the roadways and sidewalks. Using stainless steel, or most other metals for that matter, in these regions requires close attention, because the chloride ion is the most pervasive and damaging element faced by metals in the normal environments of today. Its presence in water forms an electrolyte that can aggravate the development of small pits in the surface of metal. The chloride ion can make an autocatalytic event on the surface of stainless steel, causing the pits to continue to grow on their own.

Newly polished stainless steel is very clean. Clean metallic surfaces are very reactive and will absorb water vapor and atmospheric contaminants more readily than a contaminated surface. This is why newly finished stainless steel fingerprints without hesitation. As the clean surface is exposed to the air it absorbs moisture. Tests on the surface of clean stainless steel exposed to the atmosphere have found the presence of absorbed carbon and oxygen.[1] Surface analysis showed that the stainless steel had quickly absorbed 10–20% carbon, most likely from airborne hydrocarbons and carbon dioxide. Oxygen levels were between 30 and 50%. These were instantaneous oxide formations on the surface along with absorbed water vapor.

Over time and exposure, the surface of the stainless steel absorbs H_2O. At first the oxide grows laterally to form a uniform chromium-rich oxide layer. As time and exposure to the air continues, the surface of stainless becomes multilayered with absorbed oxygen and oxyhydroxide. This

[1]Kerber, J. and Tverberg, J. (2000). Stainless steel surface analysis. *Advanced Materials and Processes* 158(5): 33–36.

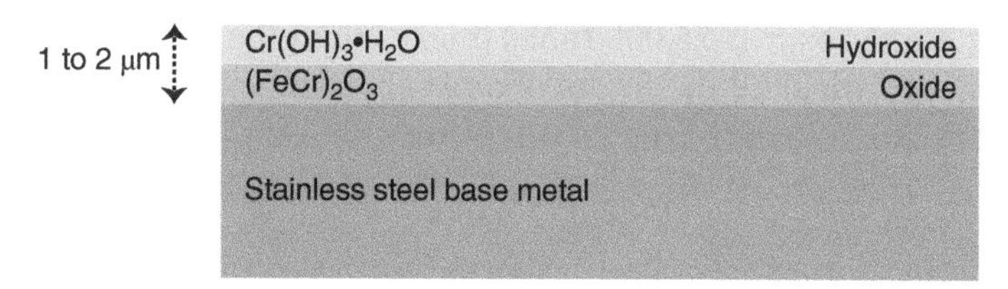

FIGURE 7.2 Passive layer on stainless steel.

thin layer is an amorphous oxide,[2] meaning the surface at the air-to-metal interface is not crystalline as the base metal is.

PASSIVITY

The corrosion resistance of stainless steel is achieved by the development of a microscopically thin layer of a passive oxide. Unlike gold or platinum, two metals that are thermodynamically passive because of their atomic structure, stainless steels are passive only because of this thin oxide layer. This oxide film is amorphous and consists of oxides and hydroxides of both chromium and iron. A passive oxide layer resists chemical reaction generated from agents in the surrounding environment.

Stainless steels are considered passive in a particular environment if anodic polarization causes the surface to resist corrosion when exposed to that environment. In other words, the surface becomes more noble, or positive, with respect to the local environment.

In stainless steels an extremely thin layer of corrosion product, in the form of chromium oxide, spontaneously forms on holes, edges, and all exposed surfaces. This is a diffusion barrier of oxide products formed with the base metal constituents. This diffusion layer separates the base stainless steel material from the surrounding environment. It is very thin and transparent. Perforated edges, reverse sides, and all surfaces when exposed to air spontaneously form this diffusion barrier of oxide. This protective layer covers every exposed part of the metal surface and is made chiefly from chromium oxide (Cr_2O_3).

In moist, temperate environments the stainless steel also will adsorb oxygen and water. A layer of hydrated oxide will develop into a dual layer composed of an inner layer of oxide and an outer layer of hydroxide. In dry, arid environments the hydroxide layer does not develop (Figure 7.2).

Chromium is the key element in creating the corrosion resistance of stainless steel. Tests on steels with more than 10.5% chromium have shown they exhibit corrosion resistance when exposed to atmospheric environments.[3] The thickness of this layer is found in the range of 10–30 Å.[4] To put

[2]Okamoto, G. and Shibata, T. (1965). Desorption of tritiated bound-water from the passive film formed on stainless steels. *Nature* 206: 1350.

[3]Johnson, M.J. and Pavlic, P.J.. (1980). Atmospheric corrosion of stainless steel. In: *Atmospheric Corrosion* (Allegheny Ludlum Steel Corporation Research Center Report), 461.

[4]Å = ångström, a unit of measure equal to one ten-billionth of a meter, or 0.1 nm. For example, an atom of chlorine is 1 Å in diameter.

this in perspective, a human hair is 25,000 times thicker than the oxide film. This very thin, clear oxide film is sufficient to withstand the forces of most environmental exposures. Only the dreaded chloride ion has the ability to cause havoc on this otherwise sound and nonreactive film.

THE CHLORIDE ION

The chloride ion (Cl^-) is a negatively charged ion of chlorine, that is, it is chlorine with an extra electron. This added electron makes it significantly larger than the chlorine atom. The sea is made up of approximately 2% chloride ions. The chloride ion has the ability to break down the passivity of the protective oxide film that develops on stainless steel. It is believed that chloride ions are absorbed into pores or enters through defects in the barrier film that forms on stainless steel. It is also believed that chloride ions may be adsorbed onto the metal surface, competing with the adsorption of oxygen or hydroxide. In this way chloride ions cause an increase in the hydration of metal ions into an electrolyte solution, as opposed to oxygen, which decreases the rate of metal dissolving into the solution (Figure 7.3).

What occurs with stainless steel is that the surface is not able to develop passivity in chloride solutions. There is a local breakdown of the passive layer and tiny areas of active metal develop into anodes in relation to the surrounding passive metal of the stainless steel surface. The electrical

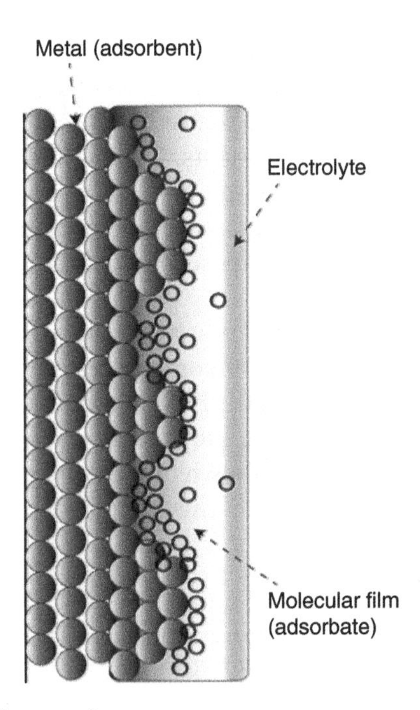

FIGURE 7.3 Absorption of chloride on surface.

potential difference, the driving force of corrosion, can be significant. This anodic spot on the surface provides cathodic protection of the area around the spot as the electrical potential develops. This causes the anode to be self-perpetuating, and the metal continues to dissolve inward, creating a pit. The ability to provide anodic protection to the surrounding metal limits the number of deep pits that may be observed on a surface.

In most instances, the stainless steel surfaces of art and architecture projects are not exposed to significant levels of chlorides in solution. Deicing salts and salt spray on a side facing the ocean are normally the vehicles for depositing chlorides onto stainless steel surfaces. Stainless steel exposed to seawater – say by immersion – will develop deep pitting. Stainless steel that has frequent contact with salt water will corrode severely where crevices hold water. Pitting will develop in austenitic stainless steels in these situations, even in the molybdenum-bearing alloys. Pitting can develop in the first one to two years of exposure if nothing is done to clean the surface.

ALLOYING ELEMENTS

Other elements are added to stainless steel to develop specific characteristics of strength and toughness. For resistance to corrosion, staining, and oxidation, the alloys of stainless steel are enhanced with several additional elements that affect the corrosion behavior under certain environments.

Molybdenum, for example, is understood to assist in stabilizing the chromium oxide layer in chloride environments. However, surface tests of S31600 stainless steel showed that no molybdenum was found on the surface. This might suggest that molybdenum reacts with active sites of corrosion, reduces the rate of a corrosion cell development, and allows for the chromium oxide layer to redevelop.

Other elements are added to aid in corrosion resistance in processes used on stainless steels. Titanium and niobium are added in small amounts to stabilize stainless steel during welding or extreme, high-temperature conditions. These elements capture the carbides before they can combine with chromium. Titanium and niobium come out of solution faster than chromium and combine with the carbon. As solidification occurs, they disperse in the alloy holding the carbon.

There is much industry debate on the efficacy of one alloy of stainless steel versus another. Substantial research and testing have been performed on various alloys of stainless in different environments around the world. It has been shown that austenitic alloys of stainless steel – in particular S30400 and S31600, the most common of the stainless steel alloys used in art and architecture – will undergo surface corrosion when exposed to chloride environments, such as seaside exposures and deicing salt exposures. Anecdotal evidence, however, suggests that once the surface is stabilized[5] and all contamination is removed, consistent and regular cleaning and rinsing of the surface with clean water will keep the metal performing very well without the development of stains and pitting

---

[5]"Stabilized" means that the welds are passive and the free iron is removed from the surface of the metal. The meaning of the word as used here differs from that in the description of stabilized alloys that incorporate titanium and niobium.

corrosion. This requires that there are no crevices where chlorides can be entrapped, or moisture will collect and stagnate. If deep pitting has occurred, removal of chlorides in the pits can be especially difficult. More than a simple wash down will be required.

ENVIRONMENTAL EXPOSURES

The Australian Stainless Steel Development Association (ASSDA) has established recommended cleaning protocols for stainless steel surfaces in different exposures and different design conditions. These have been broken down into two environmental exposures – rural and coastal – and further subdivided into sheltered and exposed areas (Table 7.1).

The use of deionized water as a rinse is recommended. Deionized water, sometimes referred to as DI water, will effectively capture unbound cations, such as calcium and iron, on the surface as well as anions, such as chlorides and sulfates. Some tap waters may have certain compounds (such as calcium) dissolved in them that can lead to streaking on the surface. Tap water contains ions and they will not draw the cations and anions from the surface.

The International Standards Organization (ISO) has defined five distinctive environmental exposure categories. These categories help to pinpoint which alloys are most suitable for a particular exposure. The categories are shown in Table 7.2.

TABLE 7.1 Recommended cleaning in various environmental exposures.

Exposure	Condition	Cleaning protocol
Rural	Exposed to natural rains	Once a year
Rural	Sheltered regions	1–2 times a year
Coastal	Exposed to natural rains	1–2 times a year
Coastal	Sheltered regions	3–12 times a year

TABLE 7.2 ISO environmental exposure classifications.

Category	Environment description
C1	Indoor environment; arid desert environment
C2	Indoor environment, unheated; rural environment
C3	Indoor environment, high humidity; northern coastal, clean urban environments
C4	Indoor environment, chemical exposure; polluted coastal, urban north environments
C5	Industrial; high humidity, high pollution, high salt-exposure environments

These classifications help to provide a direction for both the selection of the alloy and the approach to design and maintenance. The most common alloys used in art and architecture, S30400 and S31600, along with their low carbon sisters, S30403 and S31603, are suitable for C1 and C2 conditions.

Moving to C3 conditions, alloy S31600 and S31603 should be considered. As you move into the C4 conditions and C5 conditions, a designer may want to consider one of the specialty alloys or a cleaning protocol that washes the surface several times a year (Table 7.3).

Some environments and some designs can make the process of regular cleaning very difficult and expensive. In this case, it is wise to delve deeper into the stainless steel alloys that can provide the most effective resistance to corrosion for the longest periods of time between maintenance cleaning operations. Additionally, design considerations such as surface roughness, low-slope conditions, and adjacent materials should undergo a thorough evaluation to insure they will not contribute to deterioration or staining of the stainless steel surface.

The duplex alloys S31803 and S32205 have shown superior resistance to chloride attacks in environments where chloride ions are prevalent. Coastal regions where the humidity is high and northern climates where deicing salts are in common use are the most challenging environments as it relates to corrosion of the stainless steel surface.

TABLE 7.3 Suggested alloys for various environmental exposures.

Environmental exposure	Prevalent corrodant	Suggested alloys	Cleaning regularity
Indoor	Organic, cleaning agents	S30200, S30400, S31600, S30403, S31603	As needed
Dry desert	Erosion, sand	S30200, S30400, S31600, S30403, S31603	As needed
Clean; rural	Fertilizer	S30200, S30400, S31600, S30403, S31603	Rinse surface at least once every 5 years if possible
Clean; coastal, low salt, urban	Chlorides, hydrocarbon, sulfur	S30200, S30400, S31600, S30403, S31603	Once a year to remove chlorides and pollutants from the surface
Polluted; coastal, low salt	Chlorides, hydrocarbon	S31600, S31603, S32205	Once a year to remove chlorides and pollutants from the surface
Humid; coastal, high salt	Chlorides	S31600, S31603, S32205, S31254	At least once a year
Polluted; urban, deicing salts	Chlorides, hydrocarbon	S31600, S31603, S32205, S31254	At least once a year, particularly in spring once temperatures warm up

The duplex alloys became possible as modern steelmaking techniques came into practice. These alloys were developed in the late 1970s. New techniques of steel production allowed for the reduction of carbon to levels below 0.03%, making these alloys commercially practical. Unlike austenitic alloys, duplex alloys contain 50% ferrite; thus the name duplex because of the 50–50 relationship of austenite and ferrite.

The duplex alloys contain a higher level of chromium, usually 22%, and only 5% nickel. They can be welded, and their low carbon content eliminates the concern of carbide precipitation along the grain boundaries. The duplex alloys have higher tensile strength than austenitic stainless steels and forming with these alloys will pose a challenge to conventional thinking when working with stainless steels.

CORROSION

Dr. Herbert H. Uhlig, the renowned chemist and corrosion expert, defined corrosion as the destruction of a metal by chemical or electrochemical reaction with its environment. We are familiar with rusting steel and the characteristic crumbly residue as it corrodes. We are familiar with the patina on copper as it changes colors and forms a thick coating of oxide particles. These are the chemical and electrochemical reactions Uhlig refers to, as metals interact with the environment and combine with the atmosphere.

Corrosion is the tendency of a metal to return back to its mineral state, the state it was originally in before refinement by man. It is a thermodynamic process that cannot be stopped, only slowed. Stainless steel is an engineered material in which the alloying of other metals with iron effectively slows the inevitable movement of iron to one of its many mineral forms.

There are numerous industrial conditions where stainless steel must perform in conditions not unlike the surface of Venus. These include high temperatures and corrosive chemicals that require durable containment and resistance. These challenging conditions also include those in which stainless steel is used in implants to replace damaged parts of human skeletal systems, where the metal must withstand the chemistry of our bodies as well as the constant pounding of our movements, all without decaying. Transportation systems, too, use stainless steel, where it is exposed to the rigors of the combustion engine and variable road grime, along with stress of movement and impact. But for art and architecture, the environment is more benevolent. When stainless steel corrodes in an environment where humans thrive and interact, it is usually due to either chloride ions interacting with the surface, contamination from steel, human interaction with the surface, or a combination of these factors.

There are several vehicles of corrosion that will attack stainless steels. All involve the dissolution of the protective oxide film. If the film is not capable of restoring itself, corrosion will continue and sometimes accelerate.

There are several types of corrosion faced by stainless steel (Table 7.4). Of these only a few are commonly faced by artistic and architectural uses of the metal. Granted, all of the types listed in Table 7.4 are encountered by stainless steel – or at least should be considered when the metal is

TABLE 7.4 Corrosion types.

Corrosion type	Faced by art and architecture
Uniform	Common
Galvanic corrosion	Common
Pitting	Common
Intergranular corrosion	Common
Weld corrosion	Common
Erosion corrosion	Common*
Crevice corrosion	Common
Line corrosion	Rare
Corrosion fatigue	Rare
High temperature	Rare
Stress corrosion cracking	Rare
Fretting corrosion	Rare

*Erosion in the form of human interaction; that is, scratching.

designed into a project. The reality is, however, that some of these will never be confronted in artistic and architectural uses of the metal.

For corrosion to occur on a metal, there must be an electrochemical reaction on the surface. The electrochemical reaction of corrosion begins as soon as a cell circuit is formed. A cell circuit is a movement of electrons from a negatively charged point, the cathode, to a positively charged point, the anode. This can occur in the case of dissimilar metals, where one metal has a different electro-potential from another. It also can occur on the surface of the same metal. Metals can establish localized differences in electrical charge on their surfaces when they are exposed to certain atmospheric conditions that possess the ability to establish an electrochemical reaction. Depending on the environment, the electro-potential differences can be very small and thus slowly affect corrosion behavior, or they can be significant, as in the case of pollution or salt exposure. The vehicle of corrosion can also be autogenic in nature. Once it starts it will continue until the circuit is broken (Figure 7.4).

For stainless steel in art and architecture, there are several potential conditions that can develop and enable electric current to form on the surface:

- *Galvanic corrosion.* Two metals of different electromotive potential
- *Intergranular corrosion.* Differences in alloy composition
- *Stress corrosion.* Cold worked regions versus annealed regions

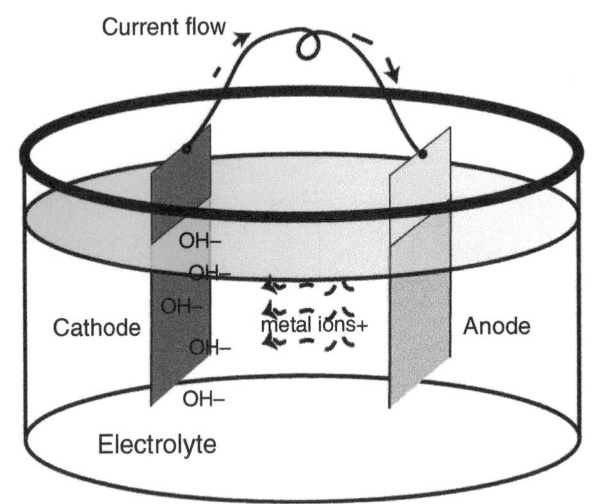

FIGURE 7.4 Corrosion cell.

- *Line corrosion.* Stainless steel element in two different solutions
- *Crevice corrosion.* Concentrations of chloride ions in crevices

For a corrosion cell to develop, one more critical element is required, and that is an electrolyte. An electrolyte is a fluid that contains charged ions. The strength of the electrolyte – how well it can carry a charge – will play a role in the corrosion rate. For dry, clean surfaces, corrosion activity will not be present. For exterior surfaces, moisture is always present in the form of condensation: as metals warm up at a slower rate than the surrounding air, their cool surfaces allow condensation to occur. When salt deposits are on the metal surface, this condensation quickly becomes a potent electrolyte.

Uniform Corrosion

Uniform corrosion happens over the entire surface or a large portion of the surface in a uniform attack on the stainless steel. From an industrial context, uniform (or general) corrosion relates to an attack on an entire surface of a metal that has been exposed to certain chemical agents. There is a measurable loss in thickness of the metal due to the overall degradation of the surface.

With respect to exposures of stainless steel in art and architecture projects, uniform corrosion could be construed to include what is referred to as "tea stains," the brown haze that develops on some stainless steel surfaces when exposed to chloride salts in high-humidity exposures. Corrosion of this type tends to be cosmetic, but it will get worse as industrial pollutants and chlorides continue to come in contact with the surface. If there are crevices or areas that remain wet, you can expect that pitting will start to develop, and the staining can deepen into reddish spots. The stain should be removed from the surface and a regular rinsing protocol should be followed each year to prevent corrosion from developing.

Tea Stain

Tea stain oxidation is the result of light iron oxide on the surface. High humidity and chloride ions from salt draw out the iron and cause it to form a slight oxide on the surface. This gives the metal a dark, sometimes spotty, sheen. It is more prevalent in coastal regions, however: after the winter season in the northern climates, deicing salts will cause a formation of a darkening stain. In the tropics, where humidity is high, the stains can form quickly, particularly in coastal areas. Stains will be more apparent on the underside of stainless steel ledges and rails, where the salts collect and concentrate (Figure 7.5).

The coarser the finish surface, the more salts are held on the surface and the greater the tendency to stain. Removing the stain requires treatment with mild acids along with a surfactant to facilitate lifting and washing the surface of the oxidation. There are several commercial stainless steel cleaners that work well for removing this stubborn stain. The best cleaners work to remove the stain by dissolving the free iron on the surface. Phosphoric and citric acid solutions with a surfactant are the most effective and environmentally friendly. Consider two passes with any cleaner to insure all free iron and salts are removed from the surface. Rinse regularly with fresh water to remove the collection of salts on the surface and keep the tea stain from returning.

FIGURE 7.5 A light tea stain forming on stainless steel.

Galvanic Corrosion

In exterior art and architecture applications, we tend to focus heavily on dissimilar metals and the potential effect of galvanic corrosion. In most applications, stainless steel is the more noble of the metals used. The problem occurs when the stainless becomes active and it loses its passivity. Even when active, however, it is still more noble than many common metals used in art and architecture (Table 7.5).

This doesn't necessarily mean that stainless steel is going to cause an aluminum or zinc element in proximity to it to deteriorate and crumble away. It does mean the aluminum and zinc will become anodic in relationship to the stainless steel and act as a source of electrons, thus sacrificially protecting the stainless steel. If the stainless steel mass is relatively small compared to the zinc or

TABLE 7.5 Electromotive scale.

Anodic polarity	Metal	Voltage potential
Least noble/more active	Zinc	−1.03
	Aluminum (A3000 alloys)	−0.79
	Steel	−0.61
	Cast Iron	−0.61
	Copper	−0.36
	50/50 Lead/tin solder	−0.33
	Brass: high zinc	−0.31
	Brass: aluminum brass	−0.31
	Bronze: silicon bronze	−0.29
	Tin	−0.28
	Stainless steel (S41000)	−0.28
	Lead	−0.27
	Nickel silver	−0.27
	Monel	−0.25
	Silver	−0.12
	Stainless steel (S30400)	−0.09
	Stainless steel (S31600)	−0.05
	Titanium	−0.02
Most noble/least active	Gold	0.14
Cathodic polarity		

aluminum, the affect will be negligible. If, on the other hand, the stainless steel mass is significant in relationship to the less noble metals and an electrolyte is present, the less noble metal will eventually be consumed in the electrochemical reaction. The rate of this consumption is dependent on a number of factors, the theoretical voltage potential difference being one.

Galvanic Corrosion and the Ratio of Areas

If the cathode area is much greater than the anode area, the current density will be larger, and this potential voltage difference will become a critical factor. This is known as the ratio of areas. Galvanic corrosion is dependent on the relationship within their contact area, or more specifically, the ratio of cathode area to anode area. The larger the cathode, the greater the galvanic current, and the smaller the cathode area, the less current will flow. For example, if a stainless steel fastener is used to fix a large aluminum cross section or panel to an aluminum frame, the area of stainless steel in relationship to the area of aluminum is very small. The galvanic current will be constant between the two metals, but the corrosion per unit of area of the aluminum will be low because the current density is low. The stainless steel fastener is needed for the strength it will provide. Stainless steel is more noble than the aluminum, and in this case the area of aluminum is significant in relationship to the small fastener. The stainless steel will be protected by the aluminum anode, and the corrosion of the fastener, even when exposed to a strong electrolyte, will be negligible to nonexistent. At the same time, the aluminum will experience very little corrosion because of vast difference in mass. The opposite condition, however, in which the mass of stainless steel is large compared to the mass of a more active metal – such as having a steel fastener in a stainless steel assembly – may result in rapid corrosion being experienced by the steel.

This is Faraday's law, which describes a condition in which a given current passes between the anode and cathode in a galvanic cell at a proportional rate. If, for example, the cathode area is 10 times larger than the anodic metal area, then the current will be 10 times as great passing through the anodic metal, and corrosion of the anode will be rapid.

If the conditions are such that you are concerned about the possible detrimental effects of stainless steel being coupled with another metal, then coat the more noble metal with an electrical insulating substance. If the more active metal is coated and there is a breach in the coating and the least noble metal is exposed, then the ratio of areas goes up. There will be a small anode area to a larger cathodic region, which is the inverse of what you want to have. You want this value to be very small. By coating the cathode, the exposed surface area will be a fraction of the anode surface area, even if the coating is breached.

$$\text{Ratio of Areas} = \frac{\text{Cathode Surface Area}}{\text{Anode Surface Area}}$$

Possibly the most damaging galvanic condition that arises and that can lead to other corrosion issues is the degradation of steel substructures in contact with stainless steel in a corrosive exposure. Here the stainless steel is the more noble of the metals, and therefore it can accelerate the decay of the steel at the interface of the two materials. As the steel corrodes, it deposits iron oxide on the

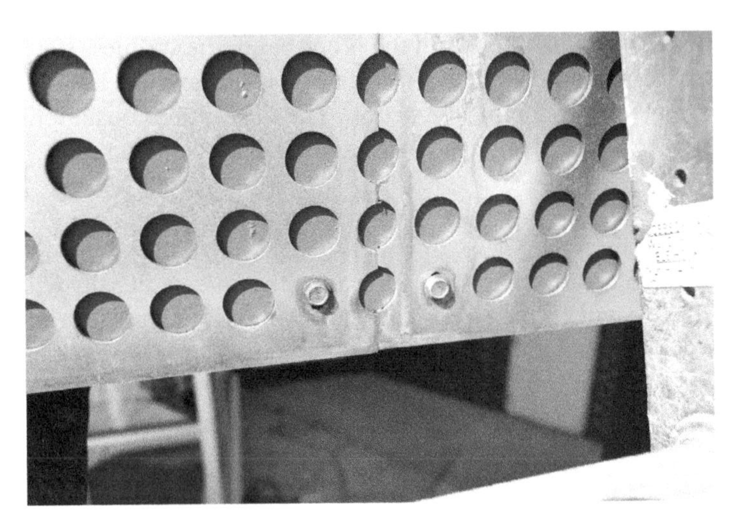

FIGURE 7.6 Corrosion of steel particles from a self-drilling fastener.

surface of the stainless steel. Over time this can interact with the stainless steel surface and create small corrosion cells known as pits. Pitting corrosion is an autogenic process that can inflict itself on stainless steel surfaces, particularly in high-chloride environments. The galvanic corrosion of one can lead to red oxide staining at best and initiate pitting corrosion of the more noble stainless steel at worst (Figure 7.6).

Key Conditions Needed for Galvanic Corrosion to Occur

For galvanic corrosion to occur, several key conditions must be met. Eliminate any one of these conditions, and galvanic corrosion will not happen. Oxygen must be available to the cathode – the more noble metal, in this case stainless steel. This usually is supplied from the electrolyte as dissolved oxygen. In the absence of oxygen, hydrogen ions form at the cathode. This can create a different issue with cathodic corrosion, but in the context of art and architecture surfaces, oxygen is usually available.

There must be an electrolyte that provides a connection for the circuit. If the environment is dry, the electrolyte is eliminated, and galvanic corrosion will not occur. If there is a barrier, such as a coating of nonconductive material or nonconductive oil, the flow within an electrolyte is prevented.

There must be electrical flow from one metal (the anode) to another (the cathode). This will take place where the two metals are in contact. This is dependent on the difference in electrical potential.

All metals have an electrical potential. The electrical potential is the tendency of a metal to give up electrons when submerged in an electrolyte. Table 7.5 shows values of different metals and is indicative of the polarity of metals in flowing seawater. The electrical potential is a measure of a particular metal submerged in an electrolyte compared to the electrical potential of a known electrode. It is an indicator "arrow" showing the electrical relationship of metals. If two metals are close to each other in electrical potential, the electrical difference should be less and thus the rate of corrosion will be slower.

Key Conditions for Galvanic Corrosion to Occur

- Oxygen available to the cathode
- Electrolyte
- Flow of electrons

Transfer Corrosion

Contamination of stainless steel surfaces from free iron and steel particles is a very common occurrence. Transfer of iron particles during the manufacturing and handling process can create galvanic cells on the surface that can lead to rust stains, or even worse, pitting. Many facilities involved with fabricating stainless steel parts and assemblies also work with other metals. For some not familiar with corrosion behavior, the idea of stainless steel having consequential damage from these other metals seems remote. However, contamination from steel particles becoming embedded in the surface of the stainless steel is a common problem. For instance, grinding carbon steel in an area near stainless steel parts can send hot, semimolten particles onto the surface of stainless steel. Even particles of carbon steel, floating aloft in steel fabrication facilities, can come to rest on the surface. Subsequent operations can embed these particles into the surface. Poor-quality tooling or damaged tooling can impart small particles into the stainless steel surface as cutting and forming processes are undertaken. Perhaps two of the worst culprits are grinding and sanding discs. These are tools commonly used in steel fabrication. If they are not isolated to stainless steel alone, or if a careful quality procedure where the discs are removed after steel fabrication is not followed, they can embed steel particles into a stainless steel surface.

During fabrication processes, usually stainless steel is protected with a thin plastic film on the face surface. This film is applied at the Mill or finishing warehouse. Steel sheet and plate are rarely protected with these films during steel fabrication processes, and steel particles can come off onto the cutting surfaces and forming dies used. If the tooling is subsequently used in stainless steel fabrication, particles can be transferred to the unprotected stainless steel surface and the edges. It is recommended that stainless steel should be fabricated on dedicated equipment and in dedicated spaces within a plant to avoid the transfer of steel to stainless steel.

Many advanced cutting tools, such as lasers, waterjets, and high definition plasma cutters, have beds made of carbon steel slats. These slats can impart steel particles into a stainless steel surface as the metal passes onto the cutting beds. In particular, laser cutting or plasma cutting of stainless steels on these beds can melt the tips of the slats during the cutting of the stainless steel and embed iron oxide particles into the surface. Often these particles go unseen as the metal is processed through the factory. These particles will have repercussions for the performance of the stainless if they are not removed. Once they are exposed to the atmosphere, the carbon steel particles corrode and can harm the protective layer, showing as red iron oxide rust on the surface (Figure 7.7).

These iron particles must be prevented from attaching to the surface or must be removed from the surface when they do. The problem with removal is that they are virtually indistinguishable from the surrounding stainless surface until oxidation occurs.

To prevent transfer corrosion, the most basic approach is to not mix carbon steel and stainless steel manufacturing processes. Grinding discs, cutting tools, cutting tables, and glass bead blasting equipment all should be devoted to stainless steel and not intermixed with carbon steel fabrication. The facility itself should have separate, distinct areas for working on carbon steel and working on stainless steel.

For lasers, cutting beds should be made of copper slats or clad with copper wraps. Copper has a different wavelength absorption rate and has been found to not contaminate stainless surface during cutting operations. They are considerably more expensive than steel and will wear, but this cost is less than the cost of a ruined stainless steel surface (Figure 7.8).

At project sites where work is underway near and above stainless steel fabrication operations, steel particles can become embedded into the stainless steel. Items such as railing and paneling, and even artwork, are often exposed to other work in process at a project site. Stainless steel needs to be isolated and protected from nearby cutting and grinding operations on steel. If the stainless steel is sprayed with these semimolten steel particles, they will embed into the surface and rust. The rust may not show up for days, and sometimes well after the work is completed. Processes to remove the embedded steel and the rust need to occur or the stainless steel finish will be permanently damaged.

Pitting Corrosion

Stainless steel is an alloy, a mixture of various elements. There are minute differences in composition across the surface of an alloy. Additionally, stainless steel is composed of crystals that have distinct boundaries between one crystal and the next. These boundaries have different electrical qualities and can create very localized corrosion cells when in the presence of an electrolyte. In most instances

FIGURE 7.7 Transfer corrosion from laser cutting.

FIGURE 7.8 Copper slats for laser cutting.

and environmental exposures this small difference in electrical potential can be ignored. Where it can become a challenge is when chlorides are present and they are allowed to remain on the surface of stainless steel. Chlorides can damage the protective oxide film and initiate a pitting corrosion condition at these grain boundaries. Elevated temperatures coupled with the presence of chlorides can have a more intensive corroding effect. Warm temperatures will accelerate the chemical reactions of corrosion (Figure 7.9).

Pitting corrosion is defined as the localized dissolution of an oxide in the presence of a solution containing certain anionic species. Chloride ions are anionic, with their added electron making them negatively charged. Pitting corrosion is sporadic. It can be localized on a surface of stainless steel but often exhibits a stochastic behavior rather than manifesting as a global surface attack. Some areas may show the development of pits while few if any develop in an adjacent area. When pitting has been initiated at a site and then cleaned of the corrosion, this is where the pitting will often return when exposure to corrosive conditions return. Most likely there is a sensitivity in the chromium oxide layer that has not fully healed.

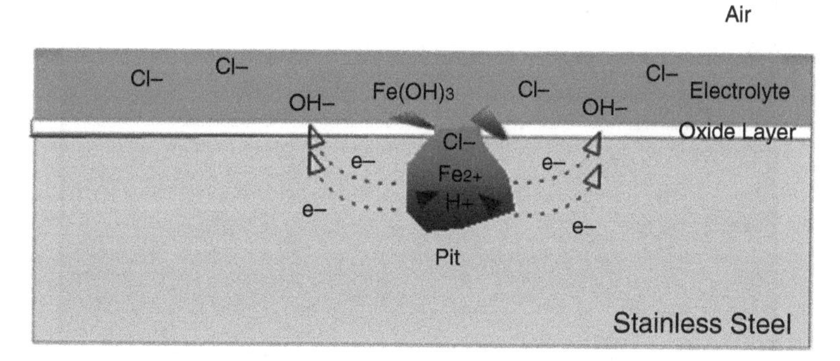

FIGURE 7.9 Pitting corrosion.

The phenomena of pitting corrosion involve a high chloride concentration leeching through a cluster of atoms at the surface of the metal and increased acidity as the metal dissolves in the pit itself. There are several theories as to why a perfectly good stainless steel surface will initiate pit development in the first place. The theories are based on both kinetics and thermodynamics and hypothesize that there is interaction between the chloride ions in the environment and the protective oxide that otherwise would keep the surface of the stainless steel passive.

Commercially produced stainless steel has varying degrees of surface inconsistencies. These are unavoidable. Things like inclusions at the granular level and distributions of composition elements within this alloy make the surface of stainless steel imperfect and somewhat heterogeneous.

Pits can be identified on the surface of stainless steel by visible deposits of dark metal around a small hole. Removal of the deposits leaves a tiny crater that penetrates into the surface of the metal. Within the pits the environment can be highly acidic. Chloride ions in the pits can form hydrochloric acid molecules when water is present (Figure 7.10).

FIGURE 7.10 Pits in stainless steel surfaces.

Pitting corrosion is a common affliction of stainless steels used near the seaside, particularly in tropical locales. The omnipresent chlorides coupled with the elevated temperature will create havoc in the surfaces of S30400 grades of stainless steel. It is recommended to use S31600, S31603, or the duplex alloy S32205. These contain molybdenum. It is not fully understood exactly how molybdenum benefits corrosion resistance in these chloride environments, but it is believed that the molybdenum aids in sustaining and rebuilding the chromium oxide layer, effectively thwarting pit development. The pitting resistance equivalent number (PREN; see Chapter 2) provides a possible guide to determining which alloy to use in particular environments, corresponding to the recommended alloys described above (Table 7.6). S32205 has one of the highest resistance ratings.

$$\text{PREN} = (\%\ \text{chromium}) + 3.3 \times (\%\ \text{molybdenum}) + 16 \times (\%\ \text{nitrogen})$$

It has been found that if small amounts (around 0.3%) of nitrogen are added to an alloy containing approximately 2.8% molybdenum, the pitting resistance increases significantly. Therefore, it is a combination of molybdenum and nitrogen that thwart the occurrence of pits.[6]

To resist pitting, consider using one of the grades of stainless steel with a high PREN coupled with an increase in surface cleaning with a neutral detergent. Consider also smoother textures, which will allow natural washing from rains. This will also aid in minimizing pitting.

TABLE 7.6 PREN values for selected alloys.

Alloy	Structure	PREN
S43000	Ferritic	18
S30400	Austenitic	19
S43932	Ferritic	19
S32001	Duplex	22
S31600	Austenitic	24
S44400	Ferritic	24
S31603	Austenitic	26
S32101	Duplex	26
S32205	Duplex	35
S31254	Austenitic	44

[6]Sedriks, A.J. (1996). *Corrosion of Stainless Steels*, 2e, 116. Wiley.

Intergranular Corrosion

Intergranular corrosion can occur with any metal made of contrastive elements. The areas around the edges of a grain that develops in a different orientation to other grains are sites where intergranular corrosion will occur. In the context of artistic and architectural uses of stainless steel, the environments the metal surface is exposed to rarely have the ability or the nature to attack. The exception is the welded stainless steel part. Here the grains are sensitive to attack due to the heat of formation and the precipitation of carbon that can occur. This type of intergranular corrosion is also referred to as weld corrosion.

Corrosion at Welds

Many processes of stainless steel product manufacturing can interfere with the development of a sound oxide barrier. Welding – or more directly the heat of welding, which liquefies the alloys around the weld – can create a phenomenon where carbon comes out of the steel and combines with chromium to form chromium carbides at the surface. When this happens, the area around the weld is depleted of chromium and thus the protective layer is retarded. The area around the weld falls below a 10.5% chromium level, which reduces corrosion resistance and makes this area more susceptible to attack. Reducing carbon content by specifying a low carbon alloy assists in the reduction of chromium carbide formation. As previously discussed, the affected zone around the weld is called the HAZ, or heat-affected zone. It is distinguishable by the darkened color that develops around the weld site (Figure 7.11).

If a significant amount of welding is to occur on stainless steel, using an alloy that has small amounts of titanium or niobium will help eliminate chromium-depleted zones. Titanium and niobium have a greater affinity to join with the carbon to form titanium carbides or niobium carbides. When these compounds form, they tend to be dispersed throughout the localized weld zone rather than accumulating at the grain boundaries as chromium carbides.

You can remove the depleted area and the chromium carbides by application of pickling pastes that dissolve the weak zones and effectively pickle the surface, allowing it to develop the passive layer naturally. You can also selectively electropolish to remove these weak areas and leave a new layer of stainless steel to form the chromium oxide protective barrier. Electropolishing avoids the use of harsh chemicals and acids (Figure 7.12).

Erosion Corrosion

Stainless steel used on architectural surfaces and artistic sculptures is not subject to the conditions that would elicit cavitation (typically, the flow of liquids and solutions moving through vessels at a high rate) as defined by erosion corrosion. Erosion of a stainless steel surface, however, can be

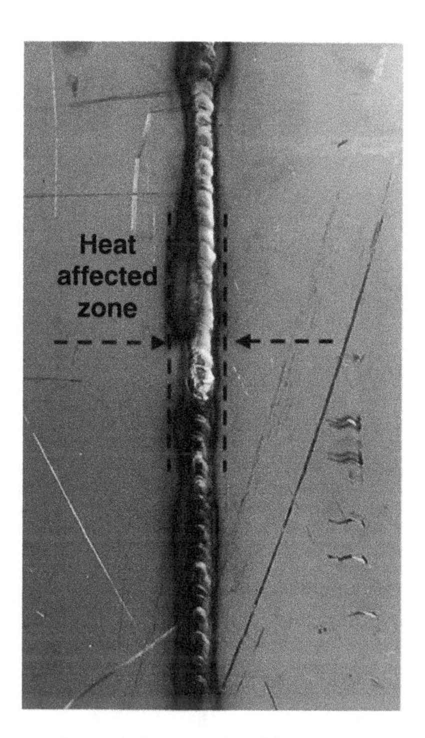

FIGURE 7.11 Heat-affected zone around a stainless steel weld.

caused by human interaction through the repeated scratching and scraping of a surface – which, even though not erosion in its truest, does occur, and does cause deterioration of the surface.

The base of the Gateway Arch in St. Louis, for instance, has had multiple scratches, deep gouges, and hammer marks imparted to its surface on the lower 3 m. The definition of erosion – the gradual destruction or diminution of something – describes what has occurred to the base of this amazing stainless steel sculpture well. If allowed to continue, it will gradually destroy the piece. It is irreversible, and in a strange way it is self-perpetuating. Others see a scratch, a name, or a mark and feel that they will leave the same (Figure 7.13).

When there is a linear finish on the stainless steel surface it is more difficult to repair, as the scratch crosses over the small peaks of the directional finish. To remove the scratch, a small amount of the metal surface must also be removed. Non-directional finishes such as Angel Hair and glass bead are actually more easily repaired. Mirror-finish surfaces can be buffed to remove scratches. In all cases there is an erosion of the surface and a finite limit to how many times this can be performed. These scratches can become sites for pitting. Steel particles can be embedded into the scratch.

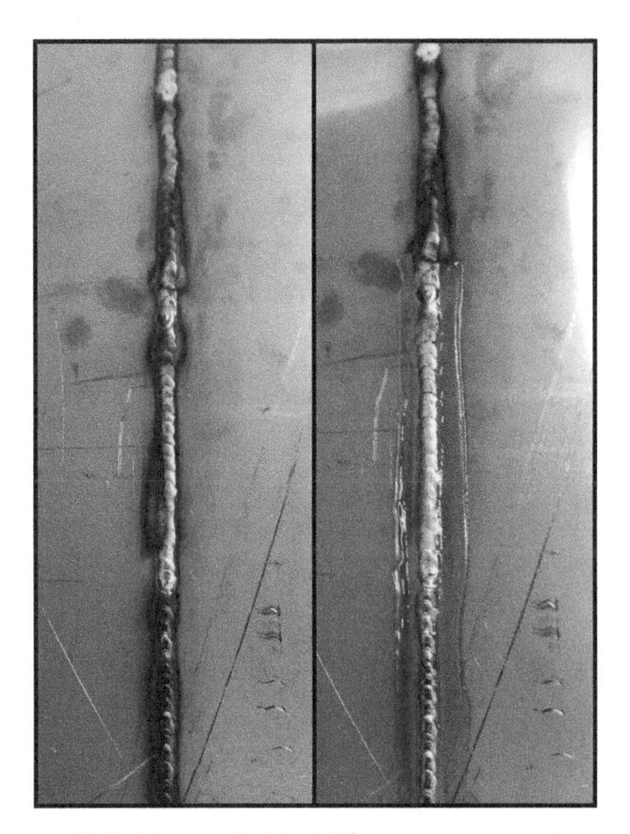

FIGURE 7.12 An untreated weld (*left*) is treated with pickling paste to remove chromium carbides (*right*).

FIGURE 7.13 Scratches in the surface of stainless steel from human interaction.

Crevice Corrosion

Stainless steel is susceptible to a corrosion behavior referred to as crevice corrosion. Crevice corrosion can occur where small gaps of less than 2 mm between the surfaces of stainless steel meet, or where surfaces of stainless steel are covered with a gasket. A gasket holding moisture or decomposing into compounds causes particular damage to the chromium oxide layer on the surface. Crevice corrosion, by its nature, is insidious. It is concealed from view. It does not extend out from the crevice and can only be uncovered by disassembly.

For crevice corrosion to occur on stainless steel there must be a concentration of chlorides in the crevice. The confinement of the crevice keeps the region wet and the active. The pH will be lower than surrounding areas as the chloride concentration increases. This can damage the chromium oxide layer in a way similar to pitting corrosion. Chloride ions can become highly concentrated in the crevice and lead to rapid corrosion.

Crevice corrosion is, in reality, a design issue. Well-designed joints can prevent the occurrence of crevice corrosion. One can eliminate the crevice by using elastic materials that allow thermal movement without bunching up, using materials that do not absorb water, and designing well-drained joints that allow rains or maintenance to flush the joints. If joints cannot be avoided, consider making them larger so they can be adequately flushed out and inspected (Figure 7.14).

Crevice corrosion is a common problem in kitchen and bathroom stainless steel cladding. This is due to cleaning fluid becoming trapped between two surfaces of stainless steel. The cleaning fluids used are often bleaches or heavily chlorinated solutions. These will enter the joints and eventually show as a light rust line on the edge. This will lead to deterioration in the concealed region.

Line Corrosion

When metals are partially immersed in one solution while parts are exposed to another solution, an electrochemical polarity can form, in particular if one solution is more oxygenated than another. For art and architecture, this might mean a fountain or basin region that retains water, or it might involve partial burial in soil. The area of interface can be more prone to corrosive attack. The more

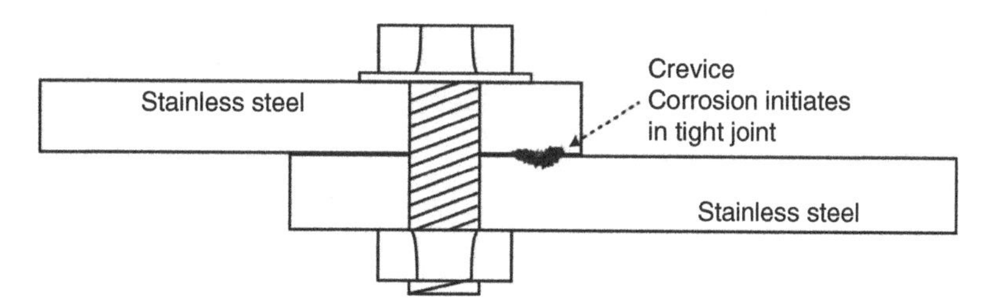

FIGURE 7.14 Crevice corrosion initiated in a tight joint.

aerated region becomes cathodic relative to the less aerated region. Oxidation, and possibly pitting, can become more prevalent at the interface region. Still there needs to be a breakdown of the chromium oxide layer for this to occur. If the stainless steel is to be partially buried or immersed in water, consider coating the stainless steel with a barrier of paint. This will eliminate the difference in polarity.

Fatigue Corrosion

This type of corrosion occurs when the metal undergoes cyclical loads – that is, loads that are applied and released or loads that place the assembly or part in tension and compression cycles. The loads are below the yield stress of the stainless steel, but still cracks may appear and propagate as the loading cycles continue. If the parts are subjected to a corrosive environment while the loading is cycled, this can create a condition known as fatigue corrosion.

This type of corrosion is not very common in the stainless steel used in art and architecture because the cyclical nature of the loads is very low and the environments are usually not so aggressive. To assist in the resistance of fatigue corrosion, use an alloy and temper that will arrive at a small grain size coupled with an increase in tensile strength. Use an alloy with high chromium, molybdenum, and nitrogen. Like crevice corrosion, fatigue corrosion is a design issue. Begin with the correct metal alloy and temper and design the stainless steel part to reduce cycle loading.

Stress Corrosion

Cold working stainless steel changes the grain structure of the stainless steel slightly in relation to unworked regions. This change makes these cold worked regions more anodic than the annealed regions on the same surface. A folded corner is going to be more anodic than the unfolded flat region. This effect is very slight, and for nearly all instances in art and architecture, inconsequential to the corrosion resistance of the metal. Like magnetic permeability in austenitic stainless steels, this change in electrical polarity is a phenomenon that should not influence the behavior of the metal in most architectural or artistic applications.

For stress corrosion to occur, there needs to be elevated temperatures in the presence of corrosive media such as chlorides and hydrogen sulfides along with an applied stress or significant residual stress. In most architectural and artistic uses of stainless steel, the conditions that lead to stress corrosion should not be present.

Fretting Corrosion

Fretting corrosion is rare in art and architecture. It is caused by two surfaces rubbing against one another. This involves the principle of asperities that touch one another at the interface of the surface. They rub against one another under vibration while loads are applied. The surfaces erode and can fail mechanically. Slip systems using Teflon pads or washers can aid in preventing fretting corrosion. Fretting corrosion can be prevented by good design and execution.

PASSIVATION

Stainless steel needs its surface to be inert, chemically inactive, and lacking the vigor to react with other elements and compounds that come in contact with it. The goal is to slow down reactions on the surface long enough to allow natural washing from rains or periodic cleaning to remove corrupting substances before they can really integrate and react with the surface. Stainless steel gets its passive, slow-to-react, ossifying nature from a thin layer of chromium oxide that rapidly develops on the metal surface when exposed to air. When exposed to the air, all sides and edges of the metal develop the passive layer. A surface that is pierced, scratched, drilled, or punched will instantaneously begin to form chromium oxide layer once exposed to air. This thin barrier of oxide is the most important feature of stainless steel. Without it, the metal is no different from basic steel with respect to corrosion.

CLEANING A CONTAMINATED SURFACE

The stainless steel used in most architectural and artistic applications will perform extremely well. With the correct alloy choice and the consistency of a cleaning regimen, it will provide a service life few other materials can compete with. The development of the chromium oxide layer and the ability of the layer to sustain itself in a passive state is what provide stainless steel the corrosion resistance that its name implies.

There are suggested treatments to help develop the passive film after working the surface of the metal. In companies that regularly work with stainless steels and are versed in the correct handling practices for stainless steel, post-passivation treatments are usually not needed. Studies of these treatments to passivate the surface indicate that these treatments are rarely necessary to enhance corrosion resistance of stainless steel in ordinary manufacturing processes.[7] On exposure to air, the surface of stainless steel immediately forms a thin oxide that offers a unique but effective barrier to the environment. It must be maintained to work.

It has been suggested that treating stainless steel with nitric acid will improve corrosion resistance. However, it has been shown that simply exposing the surface to a clean environment for 24 hours will allow the oxide film to arrive at a sufficient passive state, whether the surface is treated with nitric acid or not.

The nitric acid will, however, clean the surface and remove iron particles and other substances from the surface that may have collected during improper fabrication and handling processes. So, even though nitric acid may not be needed to build a sound protective oxide, it will leave the

[7]Wegrelius, L. and Sjöden, B. (2004). *Passivation Treatment of Stainless Steel.* Outokumpu Stainless AB.

surface clean of substances that can alter the development of a good oxide formation. From an aesthetic viewpoint, the nitric acid can reduce some of the reflectivity of stainless steel surface: the surface reflectivity will be dulled through a slight etching of the stainless steel by the acid.

Cleaning the surface with phosphoric acid or citric acid may be a safer way to arrive at a good, sound stainless steel surface without affecting the reflective quality. These acids are milder and safer to use than nitric acid. They degrade more quickly and are easier to neutralize. They do not affect the clarity and reflectivity of the surface as the nitric acid treatment will.

There are several methods of testing for passivity on stainless steel. Quick determination can be achieved by using a passivation testing device that reads the electro-potential of the surface. Measuring the voltage across any active corrosion condition allows for immediate determination of its passivity.

There are several other recognized tests for establishing the passivity of a stainless steel surface:[8]

- Water immersion test
- High-humidity test
- Salt spray test
- Copper sulfate test
- Potassium ferricyanide–nitric acid test
- Damp cloth test

These tests look for indications of free iron on the surface. The presence of free iron indicates that the chromium oxide layer is not fully developed or that there are iron particles from other sources present on the surface.

These test procedures are not easily conducted in situ on large art forms or architectural surfaces. Selected areas can be tested using either the copper sulfate or the ferricyanide test, but these other test procedures are usually not practical. An alternative method that has shown success on large art pieces and other large surfaces made of stainless steel involves a simple wet-and-dry cycle. This process will verify whether or not a surface has been cleaned of free iron and the passive layer has developed adequately. After a thorough cleaning, the entire surface is wetted with fresh, clean water and then allowed to dry. This process is then repeated several times within a 24-hour period. The temperature must be above 15°C (60°F). The surface is carefully inspected to identify any signs of red corrosion products. If any corrosion products are apparent, the surface is cleaned again using a phosphoric acid or citric acid cleaner to dissolve the iron particles. The surface is then rinsed again and the wetting and drying cycles are repeated.

It works just as well to perform the initial cleaning, allow the stainless steel surface to be exposed to natural rain and drying cycles for several weeks, then return to inspect the surface. If there is free

[8] ASTM A967/A967M-17 (n.d.). *Standard Practice for Cleaning, Descaling, and Passivation of Stainless Steel Parts, Equipment, and Systems*. American Society for Testing and Materials. doi:10.1520/A0967_A0967M-17.

iron, it will appear. Until the iron particles have dissolved or been removed completely from the surface, the metal's passive nature will not fully develop.

As with all materials, there are many internal and external circumstances that can have an effect on the performance of stainless steel. When faced with the challenges various environments pose, stainless steel performs well as a surface material. There are methods to address certain challenges and return the metal to its original condition.

CONDITIONS THAT PROMOTE CORROSION IN STAINLESS STEELS

Conditions that confront the performance of stainless steels are more detrimental because of the aesthetic personality of the metal. Anything that has an effect on the reflectivity will show more dramatically in stainless steel than in other materials. On exterior exposures there are several conditions to be avoided or addressed if stainless steel is to perform as designed:

- *Presence of contamination on the surface.* This would include carbon steel contamination from fabrication and handling and unpassivated welds. These will show as rust when exposed to the atmosphere and need to be removed in total from the surface.
- *Presence of corrosive substances.* This would include exposure to chemical fumes and acids. The most common of these comes from the use of muriatic acid and occurs when stone or concrete work is cleaned in proximity to stainless steel. The rust that appears is superficial but difficult to clean. If it is pervasive, then the surface will need to be replaced.
- *Presence of chlorides.* This would include chlorides from seaside exposures where the humidity is high and from deicing salts allowed to remain on the surface as the temperature rises. The effect of chlorides can be controlled by regular cleaning, correct alloy selection, and good design.
- *Atmospheric Pollution.* This would involve regions of high humidity and high temperature where pollution is significant. This will enable chemical reactions to occur more readily. If salts, sulfides, and other chemical contaminants are present, high humidity and high temperature will pose conditions conducive to corrosion. Regular cleaning, correct alloy selection, and good design will help fend off some of the concerns high-humidity and high-temperature polluted environments will present.
- *Surface design.* This includes areas where the surface of stainless steel is sheltered and does not receive natural washing effects. Additionally, areas where water can be trapped or where drainage will be slow and ineffective should be avoided. The lower side of surfaces where condensation will flow due to gravity and surface tension can concentrate contaminants in these areas and increase the need for cleaning regimens.
- *Surface texture.* Coarse textures can capture and hold contaminants. Rough surfaces, weld splatter, and poor welds all capture and hold contaminants and moisture more readily than smoother surfaces. Regular cleaning will be required with surface textures that are more coarse.

- *Incorrect alloy used.* A PREN of 18 is adequate for areas free of marine environmental influences. Consider alloys with a PREN of a minimum of 24 when surfaces will be exposed to seaside or heavily polluted urban environments. When chloride and pollution exposures are significant, a minimum PREN of 34 would be advised.
- *Lack of maintenance.* A basic cleaning protocol should be established. At minimum, a cleaning should be done once a year if exposure to pollution and salts is prevalent.

Having an understanding of the conditions a particular architectural or artistic surface will be confronting is critical for the long-term success of stainless steel for such exposures. Designing and planning for these conditions will produce a surface that will remain beautiful for a long period of time.

CHAPTER 8

Maintaining the Stainless Steel Surface

A gem cannot be polished without friction, nor a man perfected without trials.

Seneca

THE NEW SURFACE

Newly installed stainless steel, whether colored or natural, possesses a powerful reflective luster, as if a soft glow of light is emanating from the metal itself. This powerful appearance is unlike any other metal or material. It conveys a sense of modern, new, and clean, as well as of strength and durability. This is the distinction stainless steel carries with it and why the metal is considered for so many interior and exterior surface applications as well as prominent art forms.

The finish applied in the factory to the surface of stainless steel should be clean and unblemished. When the protective coating is peeled, the surface should be free of fingerprints, smudges, or scratches. There should be a consistency across the surface free of chatter, waviness, pitting, and other surface and reflective blemishes. The challenge is to keep it that way.

BASIC CLEANING OF A STAINLESS STEEL SURFACE

There are three categories for classifying cleanliness in a metal surface. Each of these establishes a baseline for presenting the metal and the form of the metal as they were intended. For stainless steels, this baseline is the alloying composition chosen for a particular application. Stainless steel alloys can have within them intermetallic elements designated and intended by design. Some of these will be on the surface in roughly the same levels as they are within the alloy itself. This is not an issue one needs to be concerned with except when these substances interact with the environment to introduce undesirable and malignant conditions into the balance of the metal.

Sulfur is one of those intermetallic substances added to alloying components to improve high-speed welding and machining operations. When sulfides are on the surface, they can impede finishing and affect the corrosion resistance of the surrounding metal. Iron, the main component of stainless steel, is also on the surface, but when alloyed with chromium it is protected by a thin, chromium-rich layer. Iron has an inert character as long as this protection is intact, and poses no corrosion concern.

Other substances – oxides formed when the metal is hot, grease, oils coating the surface as the metal is cold rolled, free iron from other nearby operations in the plant, fingerprints from handling, and soiling from transporting – need to be removed if the metal is to be considered clean.

There are three categories that are used to define the cleanliness of a stainless steel surface. These are:

1. Physical cleanliness
2. Chemical cleanliness
3. Mechanical cleanliness

Physical Cleanliness

Physical cleanliness is defined as a stainless steel surface that is free of grease, oils, polishing compounds, and other soils that are not related to the metal but can be found on the surface. Fingerprints would fall into this category.

Physical cleanliness is achieved by degreasing applications, which dissolve and displace these thin films that may be on the surface. Degreasing can involve solvents or solvent vapors when a completely pristine surface is desired. Physical cleanliness can be achieved by immersing a part in an electrochemical bath, which will remove organic and inorganic soils from the surface of the metal. These cleaning methods are best preformed in a plant setting, but if need be, can be performed in place using milder solvents and special controls to capture the solvents. They require proper safety equipment and procedures and proper disposal procedures must be followed to avoid all impacts to humans and the environment.

Degreasing can also involve simpler techniques that use milder, less dangerous solvents. For most artistic and architectural uses, physical cleanliness results less from the use of industrial cleaners and more from the use of biodegradable or readily accessible cleaners such as deionized water, isopropyl alcohol, and mild detergents.

Chemical Cleanliness

This relates to a stainless steel surface that is free of oxides, chlorides, nitrides, and carbides that are not related to the base metal but have formed on the metal by outside additions. These additions can lead to chemical reactions with the various alloying components on the stainless steel surface and pose long-term performance concerns.

Chemical cleaning may involve immersing the stainless steel in pickling baths to remove and dissolve oxides and scale from the surface. The pickling bath is made of nitric acid as well as other acids that can dissolve oxides that form on the surface. This returns the surface of the metal back to its original makeup and allows the stainless steel to redevelop its chromium oxide layer. This cleaning process must be preceded by physical cleaning to remove all grease and oils from the surface.

Other treatments, such as using citric acid and phosphoric acid, have shown great results in addressing mild oxides that have formed on the surface from chloride ions. Diluted ammonium fluoride solution treatments can remove mild oxides, sometimes referred to as "tea stains," as well as the more inimical spotting that can accompany advanced corrosion.

Many of these cleaning regimens can be performed in situ as well as in a plant. The handling of acids requires knowledge and expertise as well as proper disposal controls, particularly for the strong pickling acids. It is best to work with these in a very tightly controlled environment, such as a modern plant.

Advances in laser ablation techniques fall into this category of cleanliness and can be set up to achieve both chemical and physical cleanliness at the same time. Laser ablation removes all oxides and soils but does etch the base metal. Laser ablation can be performed in situ, but unless you plan to refinish the surface, it may not be the direction you want to pursue.

Mechanical Cleanliness

Mechanical cleanliness involves achieving a stainless steel surface that does not have stressed, torn, smeared, or distorted microlayers on the surface that were produced during polishing and fabrication processes, and during welding processes where the heat affected zone has developed chromium-depleted areas around the weld. This category of cleanliness is achieved by electropolishing or chemical polishing the surface to dissolve these undesirable and weakened regions on the surface.

By extension, this category can include repairing alterations created by mechanical influences on the metal, such as scratching and denting. Such repairs are more about restoring the geometric character and altered surface of the metal than removing alien substances.

ACHIEVING PHYSICAL CLEANLINESS

There are a number of soils that can find their way onto the surface of stainless steel. On certain finishes, they are not always that simple to remove. Coarse surfaces, in particular those that have been newly produced, will hold minor soils and smudges. These will interfere with light reflecting on the surface and appear as contrasting tones. The following are a list of common soils that can affect the beauty of a stainless steel surface.

Fingerprints

The first time a newly revealed stainless steel surface is touched by a bare hand a fingerprint can be seen, as the light reflecting from the surface back to the viewer is distorted. The fingerprint smudge is near ubiquitous on stainless steel surfaces. The most minor touch can leave a very thin film of human perspiration on the surface of stainless steel. This thin film interferes with the light reflecting off the surface, giving rise to a darkened smudge mark. The difficulty is that it remains on the surface until it is displaced by another fluid or removed. Fingerprinting is a challenge for many metals without clear-coating or sealing treatments, but the reflectivity of stainless steel makes the oils more apparent.

Fingerprints are composed of organic oils and fats (called lipids), amino acids, and water produced by the body. There are often salts intermixed in the oils, but the amount of chloride in a fingerprint is insignificant. The water will evaporate, but the oils and fatty substances remain. It is these oils and fats that are so persistent and difficult to move. Many cleaners just move the oils around or thin them out.

The film from our hands consists of a light organic oil, and this very thin deposit enters the minute ridges and pores on the surface. Unlike aluminum or copper, stainless steel will not be etched by these oils, but they will adhere tenaciously to the surface as they coat the microscopic ridges and depressions on the metal.

Interior stainless steel surfaces and newly installed stainless steel surfaces will show fingerprints and oils more readily than stainless steel surfaces that have been exposed to the atmosphere for a period of time. Stainless steel surfaces have an affinity for light oils. The oils penetrate deep into the surface pores of the metal. Stainless steel surfaces exposed to the environment for a period of time absorb water from the air. This water fills the pores, effectively preventing light oils from affixing to the surface. This phenomenon is more pronounced on diffuse surfaces, such as glass bead, or finely ground surfaces, such as Angel Hair and No. 4. 2D finish is dull and therefore highly susceptible to light oil deposits. Interference colored stainless steels will show fingerprinting and light oil deposits due to the oils, adding another thin film that will temporarily alter slightly the appearance produced from the thin interference film.

Oils interfere with light reflectivity from a stainless steel surface. Often they are apparent from one direction while not from another. This is a condition caused by the diffuse surface reflecting from microsurface irregularities. To remove the oils completely is not a simple task. Figure 8.1 shows a sample of No. 4 stainless steel with fingerprints and smears. This was cleaned using the following method (for directional satin finishes, always wipe in the direction of the grain):

1. Wipe with clean microfiber cloth and 99% isopropyl alcohol
2. Wipe with clean microfiber cloth and commercial glass cleaner
3. Wipe with separate, clean cloth and deionized water

FIGURE 8.1 No. 4 finish stainless steel with smears and fingerprints (*left*) and after cleaning (*right*). (The dark horizontal lines are a reflection of the ceiling beams.)

The use of 99% isopropyl alcohol breaks down the oils and smears and gets into the fine grains and displaces the oils. The glass cleaner has a surfactant that lifts and displaces the oils from the surface. Additional wiping down with deionized water will remove any detergent left from the glass cleaner, which will aid in not creating a haze. Many stainless steel surfaces look clean but when wiped down with glass cleaner they are light streaked with a filmy residue. This is created by the displaced compounds on the surface. They simply relocate.

Some commercial cleaners contain oils that displace the fingerprint oil but leave a thin film that spreads out over the surface, essentially turning a small fingerprint into a larger, oily surface. They conceal the fingerprints by filling up the adjacent surface with oil. When these cleaners dry out, they leave a streak. The typical cleaner contains petroleum distillates, which are sometimes referred to as mineral oils. These eventually will evaporate (typically after days or weeks). They can, however, collect dust from the air and hold it to the surface. If the fingerprint was not totally displaced by the oil it will reappear.

On highly polished surfaces, fingerprinting can be removed easily with commercial glass cleaners and mild soap and water. Deionized water followed by a thorough wipe down with a clean cotton cloth is an excellent way to remove light fingerprinting. Coarse surface finishes can be a little more daunting. Glass cleaners and detergent should be tried first.

Streaks and Smears

Quite often the detergent cleaners and stainless steel cleaners on the market displace substances that are in the grooves or tiny recesses of the metal's surface. As the cleaner dries, it can leave a residue of the displaced substances and soap residue. These appear as streaks and smears and alter the reflectivity enough to be visible. Streaking on the surface of stainless steel can be apparent after many cleaning exercises that use conventional glass cleaners and detergents. Denatured alcohol leaves the surface looking streaky as well. This is due to various additives that make the ethanol, the main solvent, unfit for consumption.

Isopropyl alcohol streaks less. It is usually diluted with water. It will dissolve the grease and oils, but it dries rapidly, and this can redeposit the oils to form streaks on the surface. Isopropyl alcohol of 99% purity will dissolve and displace many oils and detergents that may be on the surface of the stainless steel.

Deionized water does not leave hard-water stains and is an excellent method of removing substances; it does so through ionic attraction to the water molecule. You may have to treat the surface several times before results are satisfactory. Use only clean cotton cloths or microfiber clothes.

Another method that works for mild cleaning and fingerprint removal is to spray the surface with vinegar. Wipe the surface down with a microfiber cloth. If there is a grain, wipe in the direction of the grain. Follow with a very light wipe with mineral oil, again in the direction of the grain. The less oil the better. This removes the fingerprint and the oil fills the grain and offsets the other oils from your fingerprints.

Dirt and Grime

For exterior stainless steel surfaces, the environment begins to play a greater role. On exposure, the surface of a stainless steel can collect grime and soot from the surroundings. Vertical surfaces receive the benefit of an occasional rain to remove airborne soils that adhere lightly to the surface. Dew that collects on the surface, however, can also hold the airborne soils in place and redeposit them as streaks. An occasional rinsing of the surface with clean water and a mild detergent can remove most of these streaks. Adding the power of a pressure-washing pump can aid in removal. The surface cleaning will also benefit greatly from a final rinse with deionized water.

Adhesives

Adhesives and glues on the surface of stainless steel can be removed with solvents (Table 8.1). There are occasions when adhesive-backed plastic or paper protection is used on stainless steels. Often they can leave a sticky adhesive film on the surface. If the film is still tacky, it can be removed using solvents such as mineral spirits, acetone, or ethyl acetate. There are several proprietary solvents, some

TABLE 8.1 Selected solvents.

Hydrogen bond donor acceptor solvents
Methyl alcohol
Ethyl alcohol
Isopropyl alcohol
Ethylene glycol
Low hydrogen-bonding solvents
Naphtha
Mineral spirits
Toluene
Xylene
Hydrogen bond acceptor solvents
Ethyl acetate
Methyl ethyl ketone
Isopropyl acetate
Butyl acetate

using biodegradable citric-based substances, that can be used, but usually removing an adhesive film will require a solvent. The solvent, however, can simply make the adhesive gummy. The addition of steam can sometimes dislodge the gummy substance and allow it to be removed from the surface. Dilutes ammonium hydroxide (household ammonia) is also an effective way to remove an adhesive. Soaking the surface may be needed to loosen the adhesive. On vertical surfaces this will not be practical. If the plastic is still adhered to the surface, then you will need to start at an edge and work the solvent in to release the plastic.

When these protective films remain on the metal for a length of time they change in composition and can be very difficult to remove. They will no longer be adhesives and will have developed a crust-like hardness. This condition will require something stronger than low hydrogen-bonding solvents. To remove these films, ethyl acetate or methyl ethyl ketone (MEK) may need to be used with steam. MEK is considered a hazardous substance. Most hydrocarbon-based solvents are flammable and require special handling and disposal as well as special personal safety gear.

The use of these heavy-tack films for protecting metal are being phased out and alternative, low-tack films that are effective in protecting the metal during laser cutting operations are becoming more common. Still, it is advisable to remove the protective films as soon as practicable. Never leave them on metal exposed to ultraviolet light for more than eight weeks. After this point they begin to change and lose their resiliency. They become harder to remove and can leave a residue behind.

Deposits from Sealant Decomposition

Silicone and other sealant material are commonly used to close joints between where one material stops and the stainless steel surface begins. It is good practice to avoid the use of exposed sealants on the joints of stainless steel.

Many of these sealants undergo a catalyzing process as they cure. They contain plasticizers and oily polymers that lead to the development of stains by migrating out of the joint onto the stainless steel surface. Even after curing, the surfaces can create very adherent dirt deposits along the edge of the sealant as the dirt is redeposited around the joint. The cause of this is not fully understood and lies in the oils that exude from the sealant during curing and as it slowly decays. The oils are hydrophobic and repel moisture, so the benefits derived from rains rinsing soils from the surface is lost around sealant joints. A halo surrounding the joint will appear when the surface is wetted. This is an indication of the presence of a thin oil layer (Figure 8.2).

The sealant joints collect dirt and the dirt is redeposited onto the metal surrounding the joint. These deposits are extremely adherent stains on the metal, and due to the high reflectivity of stainless steels, their contrasting appearance has a significant esthetic affect.

Pressure washing with hot detergent alone does not remove them, nor will most mild solvents. MEK will dissolve the stain and remove the soil, but this solvent is not recommended due to the safety and environmental hazard posed by it. Ethyl acetate will aid in the removal of the stain if you allow it to sit and work on the surface. Generally, solvents tend to soften the silicone and urethanes used in sealants and allow them to be wiped from the surface.

There are a few proprietary products that will dissolve silicone. With any system used, personal protective gear along with a thorough rinse down is a requirement (Figure 8.3).

FIGURE 8.2 Sealant oils leaching onto stainless steel.

FIGURE 8.3 Cleaning sealant and other stains from interference colored stainless steel, before (*top*) and after (*bottom*).
Source: Photo courtesy of K&E Chemical Company.

***Sealant oils** leaching from sealant joints near stainless steel surfaces can create a tenacious stain that is difficult to remove. Use of solvents specially designed for sealants followed by a thorough cleaning of the surface is recommended. Do not use acids and use caution when using strong solvents that can harm the environment.*

Grease Deposits from Building Exhaust Systems

Grease that collects on stainless steel surfaces around exhaust systems used to vent kitchens or other food processing facilities can be an unsightly challenge to remove. This type of stain can be handled best by proper design and venting. The difficulty lies in the mass of the oily exhaust. Composed of carbon and fatty substances, these organic soils will alight on the stainless steel surfaces and coat the metal with a sticky crust. This greasy deposit is applied warm and can develop into a deep crust containing mold and bacteria as well as the decomposing fats.

Hot steam used under pressure with detergent can remove much of this. Deep cleaning with an organic acidic cleaner, such as acetic acid or phosphoric acid, along with a mild abrasive will aid in dissolving these tough surface deposits.

Hydrogen peroxide made into a paste with baking soda (sodium bicarbonate) and allowed to set on the surface for a few minutes can loosen these organic soils. The surface will be brighter but cleaned of the offensive material. To make the process of cleaning these areas easier, consider a more polished stainless steel. It will still appear unsightly, as the deposits will smear the reflection, but removal will be far less difficult.

High-Pressure Water Blasting

As with washing your car, high-pressure water blasting is an excellent and inexpensive method of removing substances adhering to the stainless steel surface. It is not as effective as other methods in removing oxides, but general dirt and bird debris are removed, as well as chloride salt deposits before they can have a detrimental effect on the stainless steel surface. High-pressure washing should use clean water, with or without detergent. There are proprietary additives that work well with removing chlorides from the surface as the surface is cleaned from pressure washing. A final rinse with deionized water, if possible, is another added means of removing soils and minute deposits from the surface of stainless steel.

Steam

Similarly to high-pressure washing, steam will effectively remove contaminants from the surface of stainless steel. It will not remove light oxides, such as tea stain, nor will it remove heavy oxides that accumulate around a pit. It does get deep into the metal to pull out dirt and grime from the surface. Organic contamination can be pulled from the surface, but oxides will not be touched by the steam process. The energy in the steam acts to break down the bonds that hold the grime and dirt to the surface, and the forces of the steam forces the contamination away from the stainless steel.

Degreasing Using Solvents

Degreasing with solvents is a common means of removing lubricants and oils from stainless steel surfaces. This can be as simple as a wipe down with acetone, xylene, isopropyl alcohol, methyl alcohol, or mineral spirits. These solvents do not contain chlorides, but they are flammable. They evaporate from the surface and can leave a film on the stainless steel surface. Remove the film by a final rinse with deionized water.

For the most part, architectural products used in art and architecture rarely need to be degreased. Oils and lubricants come into play if surfaces or castings undergo some level of machining. Lubricants are also used in deep drawing and some stamping operations. Factories that include these operations usually have degreasing stations in place. They often use proprietary solutions to remove the lubricants, while reducing the environmental and safety exposure these solvents introduce.

Degreasing Using Hot Alkaline Baths

Hot alkaline baths and sprays are also employed to remove grease, oils, and lubricants. Typically, hot diluted solutions of trisodium phosphate, sodium metasilicates, and other alkaline cleaners with proprietary additives are utilized. These require careful rinsing and drying operations and have limited use in parts and assemblies used in art and architecture. Hot alkaline bath degreasing would be performed in the factory making the stainless steel fabrications.

ACHIEVING CHEMICAL CLEANLINESS

Different levels of contamination require different methods to achieve chemical cleanliness. The most significant level is scale on the surface of stainless steel. Scale can be removed by mechanical means, such as blasting. But blasting can still leave contamination on the surface. Pickling in strong oxidizing acid solutions is the best way to remove heavy scale from the surface of stainless steels.

The next level of contamination relates to the heavy oxides from exposure to chloride ions. Contamination from seaside exposure to chlorides and roadside exposure to deicing salts that is left on the surface of stainless steel for several years leads to the development of deep pits and oxides on the surface. These can be cleaned by chemical treatments, selective electropolishing, pickling pastes, and laser ablation techniques. The pits cannot be sealed, but the chloride ions can be removed from the pits by special washes. Deionized water and other proprietary rinses can remove the chlorides from the pits.

Scale Removal by Pickling

Most stainless steel used in art and architecture has been pickled, so removing scale by pickling the stainless steel at the fabrication facility is rarely needed. Scale is an adherent oxide film on the surface of stainless steel. It forms on the surface during hot forming operations, which are thermal treatments such as annealing, welding, and casting. It would be extremely rare to receive stainless steel from the Mill with scale on the surface. Even thick plate is provided hot rolled, annealed, and pickled (HRAP), with the oxides from annealing removed from the surface. Scale can develop from high-temperature treatments to the stainless steel; in art and architecture it most commonly results from the high temperatures used for welding.

Heat Tint from Welding

Depending on the exposure, heat tint will not, in and of itself, damage stainless steel. If the exposure is clean water or mild industrial exposures, heat tint could remain, but it is only an aesthetic issue at this point. It does, however, represent regions on the surface where chromium is depleted and will not provide the necessary corrosion resistance if presented with a corrosive environment.

The oxides formed in welding can safely be removed by using commercial pickling pastes that contain acids or by selective electropolishing using organic acids (such as phosphoric acid) coupled with a power source. When using pickling pastes exercise caution and follow the instructions. These pastes contain various levels of nitric acid and hydrofluoric acid. After use, thoroughly neutralize the acid and rinse the surface with clean water. The weld area must first be cleaned of all flux and flux deposits, as these will interfere with the pickling. The pastes are brushed on the weld and the heat-affected region, allowed to set for a few minutes, then neutralized and rinsed from the surface. Do not allow the pastes to dry. The hydrofluoric acid in these pastes dissolves the carbide precipitation areas around the weld. The nitric acid is an oxidizing acid and aids in the development of the chromium oxide protective layer. Together they work very well in removing light scale resulting from welding operations. Some light etching can occur when the pastes are left on the surface, and these will alter reflectivity where they were applied.

Using a selective electropolishing tool, sometimes referred to as a "wand," is an effective way to rapidly remove the carbide precipitation around a weld zone and to set the surface on the way to passivation. It is safer than the pickling paste because it uses milder organic acids, such as phosphoric acid. In selective electropolishing, the weld is thoroughly cleaned to remove flux residue. The positive electrode is hooked to the part to be electropolished, making it an anode. The wand is the negative electrode, and it allows phosphoric acid to be pumped through it to the tip. The tip is a special copper electrode wrapped in a porous soft membrane material capable of holding the acid fluid. As the moist tip is passed slowly over the weld area, the phosphoric acid makes the electrical connection and acts as an electrolyte, dissolving free iron and light oxide as it passes over the surface. It is recommended to perform the pass a couple of times to be certain that all the chromium carbides have been dissolved. Once complete, neutralize the area with sodium bicarbonate solution or other neutralizing solutions and follow with a thorough rinse of the area. The process is very fast and efficient and does not involve the use and disposal of more hazardous acids, such as nitric and hydrofluoric acid (Figure 8.4).

Corrosion Potential of Welds

Chromium-depleted regions – where chromium carbides form as the carbon precipitates out of solid solution and combines with the chromium – can be easily addressed by selective electropolishing or pickling pastes (Table 8.2).

Other defects that are often induced onto the surface of stainless steel welded assemblies include embedded steel from dragging the stainless over steel tables or from contaminated grinding discs and steel wire brushes. Arc strikes on the stainless steel surface as the welder begins the welding pass

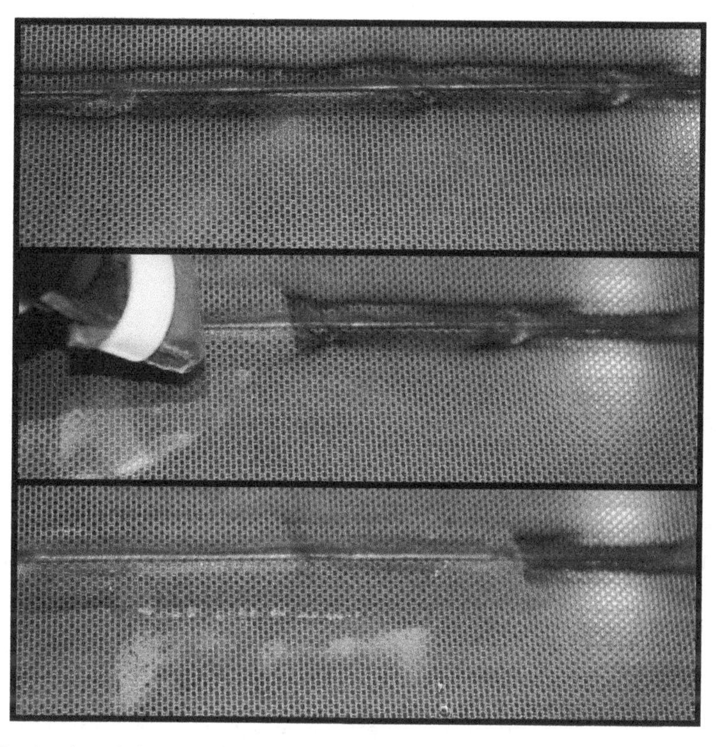

FIGURE 8.4 Heat tint (*top*) and the use of selective electropolish to remove chromium carbides (*middle, bottom*).

TABLE 8.2 Corrosion concerns from welding operations.

Welding corrosion concerns	Remedy
Chromium-depleted zones	Electropolish, pickling paste
Embedded steel	Electropolish, pickling paste
Pits from arc strikes	Grind out
Weld splatter	Grind out
Grinding marks, wire brush scratches	Finish sand
Grease pencils, marking crayons	Solvent removal
Weld flux	Stainless steel wire brush, grind out

are another source of defects. These create small pits in the surface, which can lead to the development of corrosion sites in the future. Weld splatter induced on the surrounding surface makes for a roughened area that can capture and hold contaminants. Grinding marks and wire brush scratches are coarse sites that can and will hold contaminants on the surface. Organic contamination from grease pencils and marking crayons used to scribe marks on the surface can develop into sites for corrosion if not removed.

One of the more difficult to remove substances is welding flux, which is produced by shielded electrodes and deposited along the weld bead. Grinding this away will not alone be sufficient, simply because you probably will not remove it entirely. This flux can develop into a corrosive cell in certain exposures.

Heavy Scale

Heavy scale is more difficult to remove. Often mechanical means are first employed to remove thick scale; otherwise, powerful acids capable of dissolving the base metal may need to be employed. The handling and use of such pickling acids require special safety apparatus and skill levels to protect the applicator and the environment. These pickling treatments involve oxidizing acids, such as nitric acid and hydrofluoric acid, in different mixtures. They may also need to utilize reducing acids, such as hydrochloric acid and sulfuric acid. If the conditions are so significant that they require these acid treatments, cleaning is best left to experts skilled in working with these highly hazardous chemicals. This work is not performed in place but in a plant, a facility set up to handle and dispose of these hazardous materials.

ASTM-A380-88, titled *Standard Practice for Cleaning and Descaling Stainless Steel Parts, Equipment and Systems* describes eight different methods of using nitric acid for cleaning and passivation of stainless steel surfaces and four additional methods of using it to remove heavy scale. This specification designates what a company must use in order to meet the recognized standard.

Heavy Oxidation from Chloride Ion Exposure

There are different levels of oxidation that form from exposure to chlorides. These chlorides come from exposures near the sea and from exposures near roads and sidewalks in northern climates where the use of deicing salts is common. In addition to chlorides, moisture must be present. High-humidity areas near the sea are especially challenging. The humidity develops the necessary electrolyte as moisture condenses on the surface of the stainless steel (Figure 8.5).

In northern climates where deicing salts are used, it is the warming of the metal in springtime that creates the electrolyte. Winter temperatures and the normal dry air make any reaction to the chlorides negligible. The deposits stick to the surface of the stainless steel or are caught in recesses, only to appear as rust as warmer spring temperatures arrive, along with the higher humidity the warm air brings.

The underside of surfaces is the side more intensely attacked by chlorides. This is due to gravity pulling the condensing moisture to the lower region. As the moisture moves to the lower regions, it pulls with it the chlorides on the surface and redeposits them in heavier concentrations (Figure 8.6).

FIGURE 8.5 Light corrosion on 316L stainless steel from seaside exposure.

FIGURE 8.6 Corrosion from seaside exposure (*left*) and subsequent cleaning (*right*).

TABLE 8.3 Typical causes of staining and their sources.

Cause	Source
Chloride salt deposits	Airborne salts from the sea are deposited on surfaces then become wet from humidity
Deicing salts	Salt used on roads and walks become airborne and are deposited on surfaces
Free iron particles	Contamination from cutting steel in proximity and iron dust particles from surrounding construction are deposited on surfaces
Hydrochloric acid	Acid fumes or liquid from cleaning stone or brick surfaces in proximity to stainless are deposited on surfaces
Sulfur and nitrates	Concentrations from fertilizers or industrial pollution sources are allowed to affix to surfaces

Spotting and Tea Stain Removal

Spots of dark gray and brown, rust-colored deposits, and stains will form on the surface of stainless steel that has been exposed to chlorides and high humidity (Table 8.3). If free iron particles are present on the surface, wetting is all that is necessary to form spots. The major difference between spots from chloride exposure and free iron deposits is that free iron becomes deep red and usually a small dark particle is apparent in or near the center of the spot. This particle is from the steel or iron particles that have been deposited on the stainless steel surface. Sometimes they are found in a scratch as a remnant of the article that created the scratch. Other times they may be particles that have become embedded in the surface from welding splatter or the grinding of steel nearby or simply from airborne steel deposits from adjacent corroding steel surfaces or structures.

"Tea stain" is the term used to describe a condition of uniform corrosion that can occur to stainless steels in marine environments. Sometimes referred to as "rouge," this darkening of the stainless steel surface is for the most part cosmetic, but if left on the surface it can lead to further development of more significant corrosion conditions. Eventually pits will develop on the surface. The pit is a dissolving of the metal at a localized spot and is referred to as "pitting corrosion." It begins with the stain or a discolored, darkened spot. Chloride salts from the sea, when attached to the surface of stainless steel and in the presence of moisture, can damage the protective chromium oxide layer. The chloride ions interfere with and damage this layer and can lead to the formation of pits.

More pronounced on lower, more protected surfaces than upper surfaces, this stain is the result of chlorides from the salty sea air collecting on the surface and causing uniform corrosion to occur. As the surface dries, the chloride ions can concentrate as the electrolyte carrying the ions shrinks in size. As wetting reoccurs, the chemical action begins anew, but now with a higher concentration. This is why you often see corrosion developing on the bottom edges of a surface, where gravity has pulled moisture to the lowest area. Surface tension holds it in place while the concentrated electrolyte works on the metal surface.

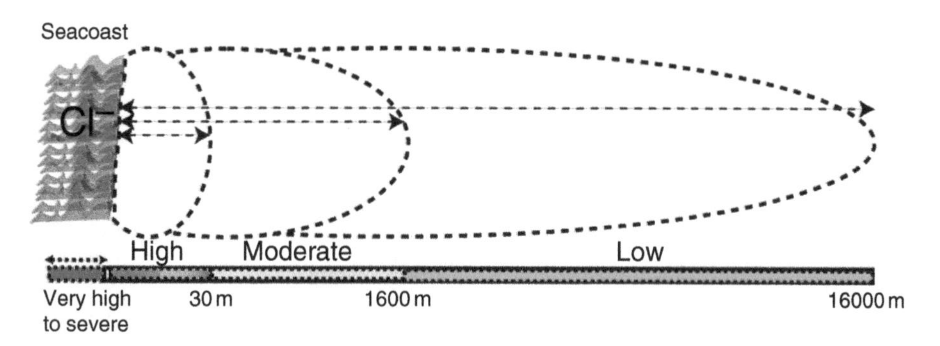

FIGURE 8.7 The intensity of chloride ions decrease with increasing distance from the seacoast.

The International Molybdenum Association has five classifications of exposure to be used when identifying the probability of accelerated surface corrosion caused from chloride exposure from the sea. Figure 8.7 shows that the intensity of chloride ions decreases with increasing distance from the seacoast. As the distance increases, the chloride ions dissipate and become less of a concern in terms of corrosive attack. However, low and moderate regions can be considered higher in probability of accelerated surface corrosion if surfaces stay wet due to humidity and shading. Sheltered regions that do not dry out and those experiencing humidity of greater than 50% are subject to an elevated rate of corrosion.

Understanding the impact of humidity and chloride ions can be critical in understanding the long-term performance of stainless steels. For example, two projects using S31603 stainless steel were constructed in close proximity to the Pacific Ocean, one in Hawaii and the other in Canada. After several years of exposure, the Canadian surface (top frame of Figure 8.8) shows virtually no corrosion, while the one in Hawaii (bottom frame of Figure 8.8) has developed the beginnings of oxidation from chloride ions. There is nothing wrong with the Hawaiian surface itself, but it does require cleaning to remove the chloride deposits creating the corrosive cells developing on the surface.

When stainless steel shows the light oxidation known as a tea stain, it needs to be cleaned and the stain removed before the stain develops into a more significant corrosion condition. These stains are from iron oxide-rich conditions that have developed on the stainless steel surface. Remove the stains using mild acid treatments involving phosphoric acid or citric acid, aided by a mild abrasive action if necessary to remove stubborn areas. These mild acid cleaners will work best if the temperature is above 20°C (70°F). A 10% citric acid solution along with 2% EDTA (ethylenediaminetetraacetic acid) applied to the surface when warm will remove the stain. The citric acid dissolves the iron oxide on the surface and the EDTA acts as a stabilizer for the iron ion in solution. Often mild abrasive action is necessary to break up the deposits that have formed on the surface. Rinse thoroughly to remove all excess acid, preferably with deionized water.

There are proprietary citric acid cleaners on the market that work to clean iron oxide stainless, as well as a few phosphoric cleaners. The ones that work the most effectively have some form of surfactant that holds the iron oxide in solution once it has been removed from the surface.

FIGURE 8.8 A project in Canada (top) and in Hawaii (bottom), both using S31603 stainless steel.

Once cleaned, the surface should have a regular washing regimen implemented to remove the salt that collects. If a washing regimen is not implemented, you can expect the corrosion to return.

Proprietary mixtures that incorporate dilute ammonium fluoride, approximately 1.5–3%, and a surfactant are very quick and effective means of removing the oxidation from the surface of stainless steels. The ammonium fluoride solution dissolves the oxidation, while the surfactant holds the sources of it and prevents them from being redeposited (Figure 8.9).

FIGURE 8.9 Staining from steel particles on a surface.

The appearance of tea stains is more prevalent on surfaces that are coarse and that are exposed to chlorides. A coarse finish can consist of the linear grain lines of a No. 4 finish or the swirls of a custom grind polish. Mirror surfaces can still develop the stains, but a mirror surface offers fewer places for microscopic particles to gain a foothold: the minute chloride ions that cause the stain tend to wash from the surface before they have an opportunity to create corrosion cells. The condition can occur on stainless steel surfaces as far away from the shoreline as 5 km. It tends to be more severe as you get near the shore.

Tea Stains from Deicing Salts

Similar stains can develop from airborne deicing salts used on roadways, and to a lesser degree, airborne sulfates or nitrates from fertilizer or industrial pollution.

In northern regions, where the use of deicing salts is prevalent, and around coastal regions, where sea salt becomes airborne and collects on surfaces, a mottled darkening can develop on stainless steel. When the weather begins to warm up, or when the cool nights give way to warming mornings, dew will collect on the metal surfaces. If salts are present, the moisture forms an electrolyte. The chloride ion present in the salt solution can attack the protective oxide layer and inhibit its redevelopment. Weak regions on the surface will begin to show signs of rust. It begins as a dark stain, usually reddish brown, and at first takes the form of a smear. It matches the wetted zones on the surface of the metal.

Deicing salts used on roadways in northern climates will create this spotting stain on the surface of stainless steel. In the United States, approximately 8 to 10 million tons of salt are spread on the roads and highways each winter.

Similar to what occurs near the sea, these deicing salts become attached to the surface of stainless steel. The arrival of warmer spring temperatures provides the moisture for the development of the electrolyte needed to initiate pitting corrosion conditions. While the temperatures are cold, the salts may collect on the stainless steel surfaces, but the electrolyte does not form until temperatures climb.

The lower portions of the stainless steel surface, particularly those that are partially sheltered, will tend to be more intensely attacked than upper, exposed regions. This is due to condensation and gravity; condensation washes the top surfaces and gravity redeposits the concentrated salts on the surface's lower regions.

If a stainless steel surface is in proximity to sidewalks and roadways that are salted in the winter, you can be assured that there will be a corrosive attack on the metal. Even if the S31603 or S30400 alloy is used, a surface will show signs of corrosive attack if salts are allowed to remain on the surface. Spring rains will reduce some of the salt deposits, washing them from the surface. The areas closest to the ground will be exposed to the chloride ion and will show signs of corrosion (Figure 8.10).

These corrosion attacks on the surface can occur quickly, in a matter of days or weeks. The coarser the surface, the more salts will become affixed, but even mirror surfaces are not immune to corrosion attacks. It is important to note that the finish on the surface is very important. Coarse surfaces, such as those produced by grinding in multiple directions, and satin linear directional finishes, particularly those perpendicular to the flow induced by gravity, will show the staining.

FIGURE 8.10 Chloride attack on a stainless steel surface.

Glass bead finishes can capture and hold the chloride ions on the surface and will require regular cleaning. Mirror polished finishes, with less microscopic texture, afford easier maintenance. They will still develop stains, but there are no indentations to hold chlorides on the surface. The mirror polished surfaces can be easily wiped off and cleaned with nothing more than deionized water.

Weaker regions on the surface of stainless steel are prone to intensive attacks. Weaker regions are areas where the chromium has been depleted or where some other contaminant is present and has damaged the chromium layer. Chromium-depleted areas are found around welds where the chromium has been depleted from carbide precipitation during the welding process. The chromium comes out of solution in the alloy and joins with carbon to form chromium carbides. The darkened color seen after welding, known as heat tint, is an indication of areas where the chromium may be depleted. Chromium-depleted zones can be removed with pickling pastes or electropolishing.

When exposed to chloride environments, these weak regions around a weld are intensely attacked. They will show as darkened, reddish rust. The importance of passivating the welds is paramount in applications where the salt exposure will be intense (Figure 8.11).

Deicing salts will have a similar effect on the stainless steel surfaces. If not cleaned, the deicing salts will darken and the stainless steel surface will develop spots that look like a disease. Eventually

FIGURE 8.11 Severe corrosion attack on stainless steel from deicing salts.

the bright silver surface will be darkened to a ruddy, reddish gray appearance. The initial sign is a darkened stain referred to as a tea stain, or rouge, that covers the lower portions of the stainless steel surface near sidewalks and roadways. As the stain forms, small, darkened circular spots are dispersed across the surface. These spots, on close examination, have a small pit forming in the center.

Initially the stain and the spots can be removed with chemical cleaners without much difficulty. But if the stains are allowed to darken and the spots become a deeper red, they can be very difficult to remove and may require several treatments (Figures 8.12 and 8.13).

Pipe sections will often show corrosion stains more readily than other cross sections. This is because of the addition of small amounts of sulfur into the alloy for the production of piping and tubing. The sulfur reduces some of the corrosion resistance of the stainless steel by developing places on the surface where corrosion cells can initiate. The pipe cross section itself tends to concentrate moisture on the underside of the form, as gravity pulls moisture, both dew and rain, to this part of the pipe form. Chloride salts that collect on the upper half are slowly washed down to the bottom half. It is very common to see the underside of a pipe or form more stained than the upper side (Figure 8.14).

The deposits of damaging chlorides decrease dramatically as the distance from roadways increases. However, studies have shown that deicing salts carried by the wind have been found as

FIGURE 8.12 Deicing salt stain, and area that has been cleaned.
Source: Photo courtesy of K&E Chemical Company.

FIGURE 8.13 Deicing stains on stainless steel sculpture before treatment (*top*), and after treatment (*bottom*).

much as 1.9 km from a roadway and as high as the 59th floor of an office building. The chloride ion is an anion, a negatively charged particle. It combines readily with other substances and thus deposits are not easily spread overland for great distances. Most likely these occurrences where chloride is spread great distances and heights involved significant winds. There have been times when a sculpture on one side of a roadway shows significant attack, while on the other side of the road a similar sculpture will show little. It is highly advised that sculptures and surfaces in close

FIGURE 8.14 Corrosion on the underside of a form.

proximity to roads have a more rigorous and consistent maintenance if they are to remain free of the stain of corrosion developing from deicing salts (Figure 8.15).

There are several chloride salts that are used for deicing. Sodium chloride is the main salt deposit in seaside environments, whereas for deicing, sodium chloride is usually mixed with other chlorides. The sodium chloride used in deicing is usually mined rather than taken from the sea. The mined salt, known by its mineral name "halide" or by its common name "rock salt," contains numerous other trace compounds. It is the most economical, yet the least efficient, for deicing purposes (Table 8.4).

Deicing salts work by lowering the freezing point of water. Sodium chloride is an endothermic salt, which means it must pull energy from the surroundings. Its effective working temperature is 7°C (20°F), so below this temperature it does little good in melting ice.

Another common salt used is calcium chloride. Calcium chloride lowers its working temperature below sodium chloride's by creating an exothermic reaction. It releases heat as it goes into solution. Calcium chloride is also hygroscopic, as it can access moisture from the air. Often deicing salt is provided in a combination of 60–80% sodium chloride and 20–40% calcium chloride.

Magnesium chloride is being used more frequently in many of the deicing mixes. This salt also reduces its effective working temperature below sodium chloride's and has less of an environmental impact because it is provided as a hydrate. This compound is composed of more than 50% water. Magnesium chloride is exothermic: as it goes into solution it will generate heat.

FIGURE 8.15 A stainless steel sign, before (*left*) and after (*right*) cleaning.
Source: Photo courtesy of K&E Chemical Company.

TABLE 8.4 Composition of seawater.

Element	Approximate % by dry weight
Chlorine (Cl)	55
Sodium (Na)	31
Sulfur (S)	8
Magnesium (Mg)	4
Calcium (Ca)	1
Potassium (K)	1

TABLE 8.5 Working temperatures for deicing salts.

Salt	Temperature
Sodium chloride	7°C (20°F)
Magnesium chloride	−18°C (0°F)
Calcium chloride	−32°C (−25°F)

Various additives have been used, many of which are organic in nature. Beet juice is a common organic additive. These additives make deicing salts sticky when they are in a slurry form, so they adhere to road and walk surfaces. They also make deicing salts stick to metal surfaces.

All of these salts corrode metals when the right conditions are met. Salts, both deicing and coastal, are hydrophilic. They absorb moisture from the air and hold onto it for a long time. Table 8.5 shows three of the most common salts used – those discussed above – and the temperature at which each tends to work to depress the freezing point of water.

There are two alternative deicing treatments that do not affect stainless steel surfaces. These are:

1. Potassium acetate
2. Calcium magnesium acetate

These are more expensive alternatives to chloride salts, but in the long run they will reduce the cost of cleaning and maintaining stainless steel as well as improve the immediate environment around roadways.

***Deicing salts** are applied in the cold of winter. The cold temperatures will inhibit and slow chemical reactions way down. When the first warm weather arrives, thoroughly rinse the stainless steel surfaces. This will remove the chlorides and salts before they have the opportunity to develop corrosion cells on the metal. Follow this simple step and the conditions created by deicing salts will be frustrated.*

Damage from Masonry Cleaners

When masonry cleaners are used in proximity to the stainless steel, they can cause stains similar to tea stains. These cleaners, which are used to wash away efflorescence, contain muriatic acid. Muriatic acid is a diluted form of hydrochloric acid, which can rapidly create an esthetic stain on the surface of stainless steel from fumes reaching the surface. If the acid itself reaches the surface, spots will quickly form. Sometimes the cleaners contain hydrogen chloride (HCl), which becomes hydrochloric acid when added to water.

When cleaning stone, precast concrete, and brick, exercise extreme caution. Overspray from these operations can have a detrimental effect on the long-term performance and appearance of stainless steels. So often these cleaners are applied via a spray and then rinsed off the surface being cleaned. Even diluted, these chemicals, if allowed to remain on the stainless steel, can have a very damaging effect. Overspray can travel far in the wind and eddy currents that wrap around a building. Cleaning the building's surface is difficult, and assurance that you have removed all of the acid when overspray happens is a gamble.

The sign that this has occurred is the rapid spotting of a stainless steel surface. The resulting small spots will turn to reddish rust spots after a few rains. They will clean off easily if caught early. They can develop small pits and corrosion cells on the surface if left (Figure 8.16).

There have been instances where stone floors in the enclosed space of an office lobby have been cleaned with acid cleaners similar to those used on masonry and fumes from these cleaners have caused stainless steel to corrode. In these cases, the stainless steel developed a reddish haze a very short time after exposure. Stainless steel rails and stainless steel electrical outlet covers several meters away were affected by the exposure. Mirror surfaces as well as satin finish surfaces were damaged. These all had to be cleaned or replaced.

FIGURE 8.16 Light spotting on stainless steel from stone cleaner overspray.

Cleaning requires some form of passivation of the surface. If discovered and cleaned quickly, the processes used to remove light oxide (tea stains) can be used to clean these surfaces. Once cleaned they will become passive again, but it is recommended to evaluate the surface after a period of several weeks to see if the staining has returned. Periods of humidity and light rain followed by days of dry warm weather will give an indication of whether the stains will return, and of the effectiveness of the cleaning process used.

Free Iron Particles

Iron particles can come from the fabrication facility, from steel dragged over the stainless steel's surface, or from nearby construction work. Most contamination seen on stainless steel comes from the fabrication facility performing work on other steel assemblies and transferring corrosion initiation to the stainless steel. If the fabrication facility is grinding steel in proximity to the stainless steel artwork or surface, you can expect contamination from minute particles of iron becoming attached to the surface.

An example of this is what happened to the National September 11 Memorial Museum in New York. The surfaces of the museum are made from S31600 alloy stainless steel with a custom polish stripe applied in the factory. There are actually several levels of polish on these panels to create a subtle blending of light reflecting from the surface. When Hurricane Sandy hit New York in 2012, it blew iron particles, oil, and other debris from adjacent construction sites onto the surface of the stainless steel. These iron particles began to corrode and stain the surface, requiring it to be cleaned. The surface was thoroughly cleaned of the free iron particles before they had an opportunity to damage the stainless steel surface.

Certain fabrication processes, such as laser cutting and plasma cutting, can contaminate a stainless steel surface by depositing molten steel particles on the surface. This can occur because most laser cutting operations use steel slats to support the sheet or plate as it is being cut. During the cutting operation the steel slats are also cut by the laser, and molten steel can become adhered to the back side or edges of the cut stainless steel surface. In laser cutting, the use of copper slat covers help prevent this. The copper has a different absorption rate, and this keeps it from vaporizing rapidly as the stainless steel is being cut.

The particles of steel that become embedded in the stainless steel surface initially appear as a roughened light splatter similar to weld splatter, but after a short exposure to the environment they develop into deep reddish streaks as they rapidly corrode. The stainless, being more noble than the steel particles, will accelerate the corrosion of the steel. These need to be removed from the surface as soon as possible. Often, in the plant, they are not readily identifiable. They will appear as small particles on the stainless steel. That they were deposited in a molten state makes them particularly difficult to remove. Often the fabricator may use mechanical means to do this. This works for some of the particles, but chemical dissolution treatments are better suited to remove all the particles. Selective electropolishing will dissolve the steel particles and render the stainless steel surface clean and able to develop the passive layer needed. In situ treatments with more specialized chemistries containing ammonium fluoride dihydrate solutions and other surfactants work well. If the parts are still in the plant, passivation treatments using citric acid with EDTA will dissolve the free iron and aid in developing the passivity. It is recommended that several cleaning applications be performed on the surface to insure removal of the iron particles. After initial cleaning, the stains and particles may appear as if they are no longer present. It is good practice to follow up with subsequent passes to be certain you have removed all the contamination. There are several tests of free iron that can be performed to verify if particles are present, but often the easiest way is to wet the surface with water, allow it to dry, then repeat this wet-dry cycle several times over a 24-hour period. If there are steel particles, they will begin to appear as small reddish spots. If they appear, then remove them; this will give a better probability that the surface has been cleaned.

Contamination from Adjacent Rusting Surfaces

In sculpture and in architecture sometimes it is desirable to use contrasting metals, such as weathering steel or copper, in close proximity to stainless steel surfaces. Transfer corrosion products made of mineral salts of the other metals may become deposited on the stainless steel. These need to be removed in order for the chromium oxide layer on the stainless steel to work effectively. Additionally, transfer rust can come from decaying internal steel structures or steel bolts used to affix a stainless steel assembly. This rust will deposit and adhere to the stainless steel surface, again, affecting the quality of the protective chromium oxide. Remove the deposits using citric and phosphoric acid treatments accompanied with a mild abrasive, using care not to scratch the surface. If this fails to remove all of the stains, a pickling paste used for weld cleanup and passivation processes on stainless steel can be used.

In sculpture fabrication and around many construction sites, steel welding and grinding is occurring. If particles – particularly hot particles – from these activities are allowed to come into contact with the stainless steel surface, they will adhere. Allowing this to occur will contaminate the stainless surface. Cleaning of the small particles must be thorough, or the rust spot will show and stain the surface of the stainless steel. These spots are initially superficial, but if allowed to remain on the surface they can weaken the chromium oxide layer and damage the metal. Use mild acid cleaners (citric and phosphoric acid) as described above. These can dissolve free iron particles. If the spray is significant you may need to replace the damaged stainless steel material. Testing the results after treatment by means of wetting and drying cycles is recommended to insure all areas of contamination have been dissolved and removed (Figure 8.17).

Free iron particles *from welding processes, grinding processes, and transfer corrosion must be immediately removed from the surface of stainless steel.*

If corrosion is severe, replacement may be necessary. Use selective electropolishing and pickling pastes to dissolve the free iron from the surface of stainless steel.

If a large artwork is transported in a steel truck bed or handled with steel forks, areas where the stainless steel surface comes in contact with the steel may be contaminated from abrasive rubbing.

FIGURE 8.17 Molten steel particles from cutting operations.

Usually affected areas appear as infrequent reddish spots where a piece of steel has embedded into the stainless steel surface (Figure 8.18).

The stain caused by steel or iron contamination on the surface differs from that caused by chloride salts, in that a small embedded piece of iron will be at the center of the spot or scratch. A stain created from iron particles will not appear immediately. In fact, it will take a few days of exposure and some humidity or rain for the particles to begin dissolving and form the stain. When this corrosion stain appears, some people mistake this for the stainless steel corroding. This stain is the iron oxide leaching out across the surface as the particle dissolves. It is a galvanic corrosion event that causes the rapid decay of the steel particle. In galvanic corrosion the less noble iron sacrifices itself to the more noble stainless steel. The minute particle size of the iron compared to the mass of stainless steel really accentuates this galvanic behavior and the particle dissolves, staining the stainless steel with a thin layer of iron oxide.

If iron particles are large, they can damage the chromium oxide protective layer and promote the development of pitting. Removal of these particles when they first start to appear is essential not only for aesthetic reasons, but for long-term performance of the surface.

Contaminated sand or abrasive discs are another source of iron particles. Often a design may call for a rough, sand-blasted texture, such as the matte finish that appears when steel is blasted by sand for paint preparation. Initially the surface has a beautiful matte, low-reflective surface appearance. But sand, even the cleanest of sand, can have small particles of iron or other substances that will, on exposure to the atmosphere, begin to show the spotting of corrosion particles. Additionally, sand blasting, unlike glass bead blasting, microscopically rips the surface and creates tiny ledges and undercuts that hold contaminants in place. These eventually develop into small pits that show as darkened spots on the surface.

FIGURE 8.18 Examples of transfer corrosion, from steel to stainless steel.

Worse conditions can develop when the sand or the abrasive discs that are used to produce a custom finish or clean down the welds are contaminated with iron from previous steel work. Fabrication facilities that do not isolate their equipment to specific metal types can use a disc that was previously used on steel or even aluminum on the surface of clean stainless and not realize that the disc has tiny particles of the other metal. The stainless steel surface will now be contaminated with these particles. The damage to a stainless steel surface by this type of contamination, caused by unknowledgeable manufacturer, can be devastating and expensive to remediate.

There are methods used to detect embedded iron particles on the surface of stainless steel. The simplest method is to wipe the area of stainless steel down with isopropyl alcohol, allow it to dry, then thoroughly wet the area with clean water. Allow the water to drain or evaporate from the surface. After 24 hours, small rust particles will be visible. Another quick and thorough method involves the use of nitric acid. Do not consider this if you lack knowledge about working with acids or chemicals, do not have the necessary protective covering, or if you are working in an area that exposes the public to harm.

For this test process, have a mixture prepared of the following:

- 94% distilled water
- 3% nitric acid
- 3% potassium ferricyanide

Spray the solution on the surface of the stainless steel and look for a blue tint to appear. The blue color is an indication that free iron particles are on the stainless steel surface. Thoroughly rinse the surface to remove the acidic solution. Note: Follow all safety precautions when handling acids. Never add water to acid but follow the acid-to-water rule. Protective eye coverings and gloves are necessary at all times when working with these chemicals.

That Profile, the Martin Puryear sculpture at the Getty Center in Los Angeles, was manufactured to the artist's design using stainless steel pipe, stainless steel cast nodes, and bronze rod. It is a fascinating example of contemporary art and was commissioned by the Getty in 1999. Soon after it was installed, the cast stainless steel nodes began to darken. The blackish stain grew across all the nodes. The nodes had a coarse, blasted surface. The first thought was that contaminated sand had been used to remove the heavy slag and oxides from the initial cast surface. Stainless steel is much more difficult to cast than a bronze or copper alloy, and the surface oxides that develop as the piece cooled are significantly more tenacious than those on copper alloy castings. However, the surface discoloration was so pervasive and unbroken across the nodes that it had to be something more than contaminated sand. It was later determined that steel shot was used at the foundry to remove the slag from the surface. Steel shot would leave thousands of tiny particles embedded in the surface.

To clean the nodes, several approaches were considered. First a mild phosphoric acid treatment coupled with a surfactant was tried. Acids, such as phosphoric and citric acid, can dissolve iron oxide if given time. This treatment, however, was not effective on this tough dark stain. Next a pickling paste containing hydrofluoric acid was tested on a small area. It did remove the dark oxide but left the stainless steel surface looking matte, as hydrofluoric acid slightly etched the surface. Additionally, applying and removing the acid treatment in situ was going to be a very hazardous process. Hydrofluoric acid is highly corrosive. It is one of the strongest organic acids known, and should only be used in extreme cases and by people expert in the handling and disposal of strong acids and bases. Even in dilute forms it will burn tissue.

The choice was to use the milder phosphoric acid with the addition of selective electropolishing (refer to Figure 8.4). The device used is one type of electropolishing device. This particular electropolishing device can be used on site. It requires that an electrical current and a way of keeping the electrode saturated in phosphoric acid be available at the surface being cleaned. Commercially similar devices are used to remove oxidation around welds, but they can also be used to dissolve iron oxide stains and the iron particles causing them. The selective electropolishing device sets up a current by attached a connection to the object and making it the anode. A cathode of graphite or other

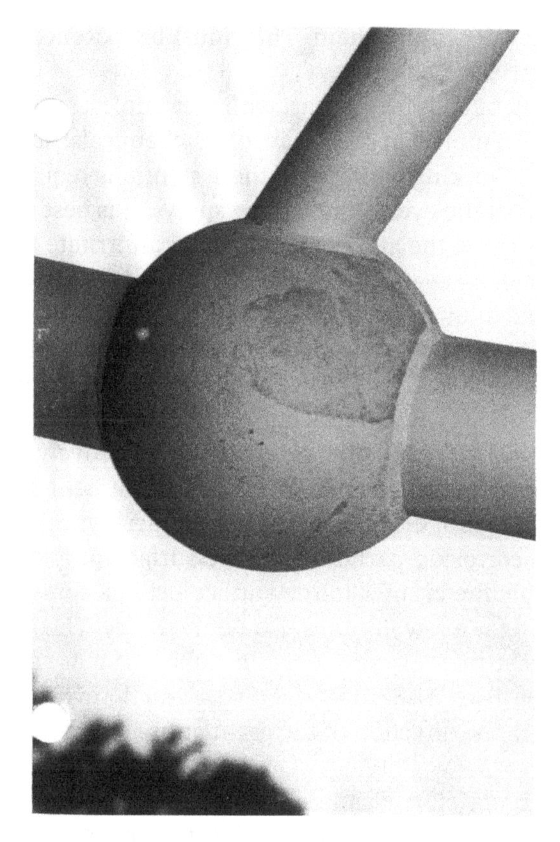

FIGURE 8.19 Oxidation on cast nodes from steel shot blast.

highly conductive material is encased in a cotton wrap. The cotton wrap is saturated with a solution of approximately 60% phosphoric acid. This acid becomes the electrolyte to make the current connection. The cathode itself never touches the piece being cleaned. The current passes through the electrolyte and dissolves the weaker oxide on the anode surface (Figure 8.19).

For the Puryear sculpture, a current was established by attaching to the base pipe near a node to be cleaned. The cathode was slowly passed over the node and the dark oxide was dissolved. This was repeated two more times to insure we were able to remove all of the iron particles embedded on the nodes. The piece was observed for two weeks with several wetting cycles. Any remaining iron showed itself as a dark red spot and these particles would be removed. After this treatment the nodes did not redevelop the oxide and remained passive and free of contamination.

Removing the Stain

The dark tea stain, transfer corrosion, and corroding steel shot on the surface of stainless steel all involve a mineral form of iron with various salts intermixed. They are very adherent, and no amount of pressure washing will remove them. They must be loosened to break their bonds with the surface.

There are several methods, but it is best to start with the simplest approach before moving on the more elaborate methods. The simplest approach involves chemicals: specifically acids, but milder forms that are biodegradable. Working with any of these solutions require skin and eye protection. They can irritate skin and harm the eyes. When using sprays, it is best to be upwind and have respiratory protection as well because the atomized solutions can irritate the lungs. Protect the areas around the work and do not allow overspray to land on adjacent materials. Have a source of fresh water available for rinsing and diluting any spills.

There are several solutions that will work to remove light stains that have developed on stainless steel. Solutions made from concentrated citric acid with a surfactant can be effective on weak stains. Stronger, phosphoric acid solutions with a surfactant are more effective, particularly on stains intermixed with spotting. Note that acid treatments will not work effectively if the temperature is below 10°C (50°F). Wait until the temperature climbs to apply the treatments.

There are a number of cleaners of the phosphoric or citric acid type that also contain a surfactant. The mild acid dissolves the corrosion particles and the surfactant allows it to be lifted from the surface and washed off. Apply the cleaner to the surface using a spray bottle and a cotton rag or sponge. Let it set on the stain for a few minutes. Do not let it dry. This should loosen and dissolve the stain. Using a mild abrasive pad very sparingly on stubborn spots can aid in the removal of the stubborn crusty corrosion particles. Use caution, however, so that you do not scratch the stainless steel with the abrasive. Rub in the direction of the grain or use a pad with a hardness below that of stainless steel.

There are very effective proprietary cleaners that use a very dilute ammonium bifluoride solution and a surfactant to lift and move oxide, oils, and dirt from a surface. These work rapidly and efficiently. They require thorough rinsing and protective personal safety gear.

Cleaning the Mirror Surface

Mirror surfaces usually come clean with a cotton rag. Just be certain there is no dust or sand on the surface that can scratch the mirror polish surface or mar the satin surface. The use of abrasives will damage the finish on mirror polish surfaces. If these surfaces are showing staining and spotting from chloride deposits, they must be handled with a bit more care. Acid treatments still will be needed to dissolve deposits. When cleaning polished surfaces, first rinse them thoroughly to remove any loose particles, grit, or sand that may be on the surface. Then soak a clean cotton cloth with the acidic solution and hold it on the surface over the area where the stain is. Allow it to remain there for a few minutes. The ambient air temperature must be over 15°C (60°F) for the process to work effectively. Lightly rub the spot and the corrosion particles should come off.

Rinse the surface thoroughly. Use deionized water or distilled water if possible. You want to be sure you get the acid residue off the surface as well as any chloride ions or free iron particles that may remain. The deionized water assists in this removal. Deionized water is water that has had most of the ionic compounds removed. The chloride ion on the surface of the metal, as well as any iron molecules or other salts, will be attracted to deionized water molecules as they seek to return to equilibrium.

Use caution when working with acids and solvents. Follow the instructions closely and wear eye and skin protection. Do not breath the atomized solutions when using a spray bottle. Have water available to dilute any spills.

Cleaning the Coarse Surface

On coarse surfaces (those surfaces that have a grain or texture), a light nylon abrasive can be used along with acid solutions to work the acid into the stain and remove it. Use an abrasive pad starting with a very soft texture first. Increase the stiffness of the pad as needed, but avoid scratching the stainless steel. Move in the direction of the polish grain as much as possible. Use a hard bristle brush, if necessary, to get into tight regions. You want to avoid altering the polish on the original surface.

Do not resort to reblasting the surface with glass bead or regrinding the surface. All you will do is embed the contamination into the stainless steel material. Stainless steel wool or stainless wire brushes will scratch the surface and change the reflective appearance, turning a bad restorable situation into a worse, nonrestorable situation. Nylon pads will usually be sufficient, but these too can scratch the stainless steel surface, so you must exercise care.

Once you remove all the spots and tea stains, thoroughly rinse the surface. Use deionized water if possible. Allow the surface to dry and then leave it for a few days or a couple of weeks. Some spots will probably return. It is very difficult to get all the contamination from the surface in the first treatment.

Return and inspect the surface. For the spots that return, treat the surface again to remove them and lightly go over the entire surface again. Rinse it down and allow it to dry. Then reinspect it to see if the contamination has been eliminated.

You can speed the inspection of the surface up by wetting the surface, allowing it to air dry, then wetting it again. Do this several times over a 24-hour period. The air temperature should be sufficiently warm to dry the piece. The spots will begin to appear during this period if contamination persists on the surface.

Once the surface has been effectively treated and cleaned, the stainless steel luster and appearance can be maintained easily by periodically cleaning the surface down with clean fresh water and light scrubbing it with a bristle brush. There are several proprietary chlorine inhibitors[1] that can be added to the wash that will help in removing chlorides that collect on the surface. These substances

---

[1]Chlor-Rid® is a product from Clor Rid International. It is added to the wash used to clean the surface down to remove salts from the surface of metals.
Chlor-X is a product from Rustbuster. It comes in various forms and can be added to the wash water to remove chloride salts from the surface.

chemically bond with the anion and allow it to be removed. Activated alumina, for example, can act as a filter to remove anions from water. Chloride washes and inhibitors chemically bond with the chloride, keeping it from affecting the chromium oxide layer.

Laser Ablation

Laser ablation is a cleaning method that can be used on stainless steel and other materials to remove oxides, chlorides, dirt, grease, paint, and grime. The development of fiber lasers has made this technique a viable form of cleaning and restoring a metal surface to its original form. Laser ablation removes material from the surface of stainless steel by thermal shock and vaporization.

Essentially, the energy of the laser light is absorbed by the surface contaminants, and this absorption of high energy vaporizes the substances without damaging the stainless steel. The laser beam is pulsed on the surface and a tremendous amount of energy is released in a short period of time. This energy is absorbed by the particles on the surface of the metal, even the oils of fingerprints. The particles become excited and vibrate as the bonds between the substances and the stainless steel surface are broken. A flowing gas sweeps the particles from the surface and suction pulls the particles into a collection filter. In this way, surface contamination and oxides are collected and disposed of without releasing them into the atmosphere or redepositing them on the ground. This also protects the lens of the laser device.

In laser ablation, the laser itself is kept remote and the beam of light is transferred via a fiber cable. The cable length can be more than several hundred feet if necessary without affecting the power at the beam-to-metal contact. The portability afforded by the fiber laser allows for cleaning remote and difficult regions in situ.

The most common fiber laser used for laser cleaning and ablation is a Nd:YAG. The Nd:YAG is more efficient than a CO_2 laser and it is easy to handle and operate. The Nd:YAG is a solid-state laser based on neodymium-doped (Nd) crystals of yttrium aluminum garnet (YAG). These are man-made crystals specifically designed for lasers. The Nd:YAG uses a doped fiber to deliver high-power energy to the laser head. For ablation cleaning, the Nd:YAG is normally used as a pulse laser, with an emission wavelength of 1064 nm. There are several alternatives to the Nd:YAG that utilize other rare-earth elements, and the development of these lasers is advancing rapidly. But the majority of ablation operations are performed with the Nd:YAG fiber system.

For laser ablation to work, the material's wavelength absorption must be compatible with the laser wavelength. The energy needed to break the bonds that hold the oxides and detritus to the surface is below the threshold of the bonds that hold the metal together. When the bonds that hold the contamination to the stainless steel are broken, the contamination vibrates off the surface from the excited, high-energy electrons.

Unlike other methods of cleaning, no waste is released to the environment, the process requires no solvents, applies no water and leaves the surface free and clean of almost all contamination. Laser ablation can be used on wet or dry surfaces (Figure 8.20).

Special training and care need to be exercised in laser ablation cleaning of stainless steel surfaces. The skill required and the expense are higher for this cleaning technique than for other

FIGURE 8.20 Laser ablation.

chemical techniques. However, the residual waste and environmental impact are significantly lower for this cleaning technique.

In using laser ablation, the surface of the stainless steel will be altered. The pulsing laser leaves behind a fine etched finish that is different from other mechanical finishes. It can be performed robotically or manually. Manual applications may leave a finish that does not meet the visual requirements and aesthetics of the original surface, making refinishing necessary.

Dry Ice Blasting to Remove Corrosion

Dry ice blasting can effectively remove light corrosion conditions that develop from deicing salts and other contaminants. Heavy corrosion will need more intensive processes such as electropolishing, laser, or chemical applications. Dry ice blasting works well in situ. On thin material, run a test first to be certain you do not deform the metal from the impact of small dry ice pellets. Dry ice blasting, also known as cryogenic blasting, effectively removes oxides, dirt, adhesives, grime, and even paint from the surface of stainless steel by impacting the surface with a high-pressure jet of the solid form of carbon dioxide. The temperature of the ice pellets is approximately −75°C, and as they hit the surface of the metal, a thermal shock occurs as the immediate localized temperature drops. This thermal shock displaces moisture and corrosion particles as the dry ice particles rapidly expand in volume and as the solid carbon dioxide turns into its gaseous form. This process is known as sublimation. The impact pressure used to blast the dry ice is provided by an air compressor. This pressure also aids in the removal of contaminants from the metal's surface.

There is no moisture or residue left behind since carbon dioxide has no liquid state. There are no acids, pastes, or solvents to dispose of. The residues removed from the surface need to be collected and disposed of, but there is no moisture created and this makes cleanup minimal. Protective gear and special delivery systems are necessary. You will need a hopper to hold the dry ice, which needs to be in proximity to the process. You will also need a compressor for providing adequate pressure. The sound is deafening, so ear protection (as well as eye protection) is required. Keeping the public protected is also necessary. It is also important to note that caution should be exercised if this process is used in a confined space. Carbon dioxide is heavier than oxygen and will displace air as the pellets return to gas.

The carbon dioxide used to make dry ice is removed from the atmosphere in the form of 3 mm pellets. The process of blasting puts the carbon dioxide back into the atmosphere, so there is no carbon dioxide created.

ACHIEVING MECHANICAL CLEANLINESS

Electropolishing

Electropolishing is a process commonly used on stainless steel. Electropolishing removes metal surface ions in a slow, controllable, and predictable manner. The process is also referred to as electrolytic polishing. Electropolishing can be performed in the plant by immersing the stainless steel part into a custom formulated acid electrolyte bath and applying a direct current. This technique can remove the trace surface particles and tiny surface irregularities left from other mechanical operations as well as surface contamination, such as oxides.

Electropolishing can be used on any of the stainless steel alloys, on any forms (wrought or cast), and on any of the mechanical finishes. It cannot be used on interference colored stainless steels or heat colored stainless steels. The electropolishing will remove a small portion of the oxide and this will remove the color. Because of this, however, it can be used to create decorative surfaces by applying a resist and removing the color from the unprotected surface.

Electropolishing involves making the stainless steel part the anode and immersing it in an electrolyte. A cathode is set into the electrolyte and a thin film forms around the surface of the stainless steel. This film is thicker over microscopic depressions on the stainless steel surface and thinner over microscopic protrusions on the surface. On the microscopic protrusions, the resistance is reduced and dissolution of metal ions on these outer regions occurs. Stainless steel contains many small microscopic intermetallic and nonmetallic particles across its surface. Iron oxide from rusting free iron, carbide precipitation, and sulfur particles can be exposed on the surface from manufacturing processes. Mechanical abrasion does not remove all of these and can actually further embed them into the surface. Electropolishing removes these from the surface and renders the stainless steel a single electropotential, with a consistent, chromium-rich layer.[2] Surface roughness is reduced, carbon is removed from the surface, and metal oxides are dissolved.

Electropolishing will clean, brighten, and level out the surface of stainless steel. The metal is immersed in an electrolyte and made anodic by running a direct current between a cathode in the

[2]Auger electron spectroscopy (AES) readings of a post-electropolished surface found that the surface had very low levels of iron in the oxidation state, a reduction of sulfide inclusions, and a reduction of precipitated carbides.

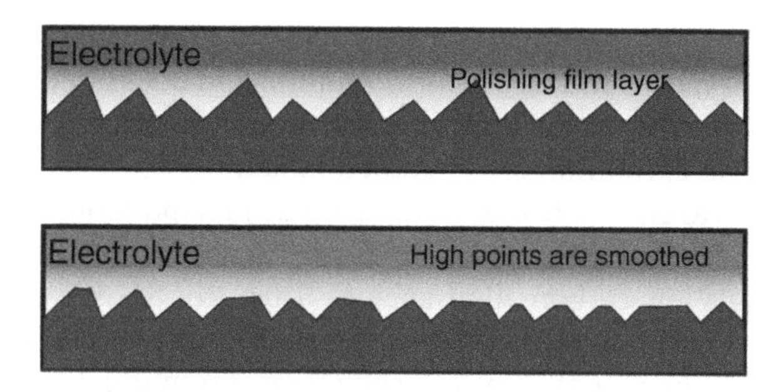

FIGURE 8.21 Stainless steel surface before (*top*) and after (*bottom*) electropolishing.

tank and the part. The peaks and irregularities are dissolved by preferential dissolution of the high points on a metal surface.

When the current is applied, a thin layer forms directly off the surface. This very thin layer is produced by the metal and the electrolyte reaction and has a greater electrical resistance than elsewhere in the electrolyte. The high points or protrusions from the surface will possess a much greater current than the minute surface depressions. This causes the metal to dissolve into the electrolyte at a much faster rate. The result is the removal of asperities and contamination that are on the surface and a level surface. The surface of the stainless steel will brighten from the removal of small irregularities. Irregularities as small as 0.01 μm are eliminated as they are dissolved into the electrolyte (Figure 8.21).

Deionized water will assist in the removal of chloride ions on the surface. Deionized water is water that has had most of the ionic compounds removed. The chloride ion on the surface of the metal, as well as any iron molecules or other salts, will be attracted to deionized water molecules as they seek to return to equilibrium.

Heat Tint Removal

Heat tint from welding operations is a common development in the fabrication process. Heat tint is a discoloration that occurs around the weld zone. It is dark in color, with an occasional blue, red, or yellow interference film that is created by the intensive heat of welding and the subsequent cooling. When stainless steel is welded, the temperatures around the weld zone are in the critical range of 600–900°C. When these temperatures are reached, chromium carbides precipitate at the grain boundaries. This weakens the corrosion character by depleting the chromium in the weld region. Carbon, a component of all steels, comes to the surface, and as it cools combines on the surface with chromium to develop chromium carbide. Chromium carbide is the dark color. These areas weaken the stainless steel at the surface around the weld. When exposed to the corrosive atmospheres, the depleted chromium zone will corrode around the weld.

When stainless steel is supplied from the steel Mill it is always in what is called a "solution annealed condition." That means the carbon is in solution and does not collect at the grain boundaries as chromium carbides. When the critical temperature is achieved, the carbon comes out of solution and combines with chromium at the grain boundaries.

There are several ways to eliminate chromium carbide precipitation and the subsequent reduction of corrosion resistance around the welds. One way to reduce this is to reduce the amount of carbon available by using a low carbon alloy, such as S30403 or S32603. Another way is to use an alloy of stainless steel that has titanium or niobium as stabilizers in the alloying mix. Titanium and niobium are found to have a greater affinity to carbon than chromium. When the carbides come out of solution during heating, they combine with the titanium or niobium. When the metal cools, the titanium or niobium carbides remain disbursed and do not collect along the grain boundaries. Alloy S32100, for example, is alloy S30400 with a small percentage of titanium added to stabilize the alloy, while alloy S34700 has small amounts of niobium.

A second way is to remove the carbides from the surface after they form. This is done by dissolution. To dissolve these chemically there are several excellent proprietary sprays and wipes that contain strong phosphoric acid with a surfactant. Other treatments use pickling pastes and gels. Pickling pastes and pickling gels contain stronger acids, such as nitric acid and hydrofluoric acid. With either approach, the point is to dissolve the carbide and remove the weakened portion of the stainless steel surface. This renders the remaining metal surface passive. Electropolishing (also called selective electropolishing and electrocleaning) is an exceptional method of removing the heat tint, carbides, and weak metal surface, leaving the stainless passive. The systems are portable and not as dangerous as pickling pastes and gels (refer to Figure 8.4).

Some fabrication facilities will simply grind the dark color away. This may remove the color, but it often still leaves the weakened, chromium-depleted stainless steel on the surface to rust later when exposed. This is not an effective way of passivation. Chemical means are more effective.

If you have a fabrication that has a significant amount of welding, it is recommended to use a low carbon form of stainless steel that is one of the stabilized grades (Table 8.6).

TABLE 8.6 Alloys with low carbon or elements supporting weld stabilization.

Alloy	Description
S30403 (304L)	Low carbon austenite
S31603 (316L)	Low carbon austenite with molybdenum
S31703 (317L)	Low carbon austenite
S32100	Titanium added to S30400
S34700	Niobium added to S30400

*In eliminating **heat tint and chromium-depleted zones** around welds, you should consider:*

- *One of the low carbon alloys, such as S30403 or S32603*
- *Alloys with titanium or niobium, such as S32100 or S34700*
- *Post-weld treatments with selective electropolishing*
- *Post-weld treatments with pickling pastes*

Weld Splatter

During fabrication that involves welding, weld splatter onto adjacent surfaces near the weld process can create aesthetic issues as well as corrosion issues. Weld splatter can create weakened regions on the stainless steel surface or areas where pollutants can be trapped and held.

If the splatter is from steel, it must be removed quickly. The surface will be contaminated and will corrode around these spots and render the surface nonpassive. If the splatter is stainless steel, the surface will be darkened around the splatter due to the heat of these fragments. They must be removed, and the surface must be treated to return to a passive state.

To protect the stainless steel surface around welding operations, use protective blankets. For areas directly adjacent to welds, there are several proprietary treatments that can be applied to the surface that will keep the hot splatter from adhering to the surface and becoming partially embedded.

Scratches

From a cosmetic viewpoint, one of the most destructive impacts on many stainless steel surfaces comes from – unfortunately – human interaction. For some reason, human interaction with stainless steel somehow creates the desire to purposely scratch the surface. Some people find it rewarding to carve their initials into a stainless steel surface. Maybe it's the perceived permanence of their name or initials carved into the metal (Figure 8.22).

Removal of the scratches is daunting. The first step is to thoroughly clean the surface. Remove any steel particles that may be deposited in the scratch from the device used to make the scratch. Once cleaned the polish can be restored.

Repairing the Damaged Stainless Steel Surface

One of the easiest finishes to repair is the mirror finish. Light scratches can be polished and buffed to match the surrounding surface. Even heavy damage can be polished to a point where the damage is reduced significantly. Since this finish was produced by polishing and then buffing, returning

FIGURE 8.22 Scratches and oxide stains on a stainless steel surface.

its luster is simply a matter of skill and effort. When viewed at different angles, a mirror polished surface can show very slight, tiny polish lines. These grit lines, very minute curved halos, are the result of very fine abrasives that produce the final polish. Some mirror-like surfaces have very few, if any, visible to the eye. These are very fine rouges that were applied during the end buffing process. Others have numerous tight swirls that become visible when viewed at oblique angles (Figure 8.23).

On mill finishes, on the other hand, it is nearly impossible to repair surface damage. These finishes are induced into the surface as the metal passes through rolls at the Mill source of the sheet or plate. If the surface area of the work is small, it is better to produce a finish over this entire surface that resembles the original mill finish. This is easier said than done. Spot repairs on the surface can make the surface appearance worse due to light reflections off the metal.

Between the mill finishes and the highly polished finishes there are numerous other finishes with varying degrees of difficulty in repair. In all circumstances, consider a sample test performed on a similar surface. There is a bit of skill involved, and you will want to hone the technique to ensure the repair work does not make things worse. Repairs to mechanical finishes require stages and the ability to "feather out" the finish to match the undamaged regions. Before you begin the repairs, you must determine the nature of the finish you are trying to restore and the degree to which restoration is feasible.

Mechanical finishes fall into three categories:

1. Blast-applied finishes
2. Linear satin finishes
3. Non-linear satin finishes

FIGURE 8.23 Visible grit lines on polished stainless steel.

Blast finishes produced by the high-energy impact of glass beads or stainless steel shot produce a non-directional medium-to-low gloss surface. When these finishes are scratched or marred, they can be repaired by first polishing out the scratch and the area around the scratch. Using a small compressor, hand blast the polished area and feather out onto the region being matched. The blast media must be matched in grit size and air pressure. Apply the finish as one would apply paint, moving across the area being repaired and not allowing a concentrated stream on any given area. You must polish out the surface first and take it to a mirror-like reflection before applying the glass bead blast or shot blast surface.

Linear satin finishes can be repaired in a similar manner. First polish out the scratch or mar. This requires sanding down with a 150–180 grit first, matching the direction of the grain of the surrounding area, then stepping the grit up to match the grit on the final piece. Allow the work to cool between sanding operations to avoid heat buildup. Since these finishes were machine applied, the difficulty arises in three areas. First, you need to identify a final grit size closest to the surrounding finish. This may involve tests and measurements of roughness, but it will definitely involve visual review of the reflective quality of the surface. If you do not know how the linear satin finish or the No. 3, No. 4, or Hairline finish was produced, then you will need to experiment with different grits and with different media. Aluminum oxide produces a different cut pattern than silicon carbide. Scotch-Brite produces a different pattern as well.

The second problem to overcome involves the mechanics of applying the finish.

Was the finish applied with a hand-finishing belt or machine applied by a CNC or semiautomated system? You can determine this by examining the finish. If there is a slight curvature or definitive overlaps, it was hand applied. If the surface is consistent and parallel, then it was applied via a fixed machine (Figure 8.24).

If it is determined that the finish is a fixed machine-applied surface, then the directional polish must be applied with a fixed machine or blocked in such a way that the hand application matches the parallel nature of the machine-applied finish.

The third issue to overcome is the start and stop points of the finish. If you go edge to edge, the potential for the edge to round off exists due to the pressure of applying the finish. The same occurs when removing the pressure. A soft line can be induced as the belt is raised from the surface.

Other satin finishes, such as the Angel Hair finish, require similar care, but to a lesser degree. One will need to match the grit of the final sanding polish and approximate the arching direction. Some feathering out will be needed to blend the finish to match the surrounding areas. Angel Hair and Vibration finishes are easier to blend than their straight-line cousins. The surface is thoroughly cleaned, and the grit used to produce the original finish is matched. Feathering in is done with the hand-applied tool to reduce the development of a depression and to blend any major damage.

FIGURE 8.24 Applying a No. 4 directional finish to a damaged surface.

FIGURE 8.25 Embossed patterns.

Etched surfaces created by dissolving the surface to create a matte finish fall in the same category as the mill surface finishes. If you are trying to repair the part of the surface that was removed by acid etching, then one could concentrate only on the etched section and apply a different finish.

Rolled-on and coined finishes are similar to mill finishes in that they are also created by passing the stainless through patterned rolls. Some of the deeper patterns were created to help conceal damage and scratches (Figure 8.25).

Blasting with sand, steel shot, garnet, or other media can create corrosion problems. Steel shot will damage the surface and embed iron into the surface. Even clean sand can introduce contaminates into the stainless steel surface. Garnet, like sand, will render the surface very coarse – so coarse that cleaning the surface is difficult.

When stainless steel finishes are damaged and then repaired, there is a localized change in the light reflection from the surface. In certain light conditions the damage or change may not be visible, while in other conditions it becomes readily apparent. This masking and unmasking have to do with the scattering of light and the angle the surface is viewed from. In bright daylight, stainless steels reflect the blue range of the spectrum. In overcast conditions, the light is highly scattered, and the blue portion of the spectrum is subdued. Conditions on the surface of stainless steel may not be visible under bright sunlight but then appear during overcast periods. The opposite effect can also occur depending on the surface reflection characteristics.

Ghosting from Packaging Materials

On occasion, when stainless steel surfaces are protected and stored for periods of time, the protective media can leave light stains on the surface. In particular, art made from stainless steel that has been wrapped in plastic bubble wrap and placed into storage is subject to a surface mark that is a ghosting of the bubble wrap. This light staining matches the pattern of the bubble wrap and is very adherent. Additionally, on sheet or panels made of stainless steel and wrapped in plastic protective sheets, very light oxidation stains can form where air is trapped between the stainless steel surface and the plastic. The film can be electrostatic and have little, if any, adhesive, yet a mark corresponding to the air bubble will remain on the surface (Figure 8.26).

These stains cannot be cleaned by conventional solvents or light detergents. They are surface changes created by trapped moisture and differential oxidation. Some additional decomposition of the plastic could create a microclimate as the temperature of the piece changes.

Viewing the surface at high magnification shows tiny microscopic trails like filiform corrosion, but reducing treatments using mild organic acids have no effect in removing them. Time and exposure do not make the stains less apparent either.

Light abrasion can be used to remove the appearance. Caution is needed about brightening the spot where the light abrasion is performed. Addition of calcium hydroxide (lime) powder assists in the blending of the spot. Selective electropolishing of the surface will also remove the stain but will brighten the region. Practice on a less visible area to see how the finish looks.

FIGURE 8.26 Ghosting from bubble wrap packaging material.

DISTORTION DUE TO FABRICATION AND WELDING

Preventing distortion in welded stainless steel can be a daunting endeavor. It is usually created at the design stage and left to the fabrication operation to overcome. For stainless steel, it is especially difficult due to the low thermal transfer of the metal. The heat is retained near the source. The surface of the stainless steel will show distortion without hesitation due to nonuniform expansion and contraction of the metal near the weld as well as the weld metal itself. Stresses develop as the metal expands around the hot weld, while the metal away from the weld remains cooler, resisting the expansion. As the weld cools it wants to shrink and return to its normal volume, but the metal adjacent resists this. The stress levels at this point approach yield and the weld metal stretches and thins out, relieving this stress as it cools. This can sometimes make the weld concave. Some of this stress relief is transferred to the surrounding metal, causing it to deflect and distort.

The shrinkage forces in the weld as it cools can induce distortion in the part. The distortions can cause bowing, rotation, and oil canning[3] of the surface.

When welding stainless steel, the use of copper chill bars and aluminum backing blocks are common practice to remove the heat more rapidly and cool the part down so that less of the surrounding metal stays hot. Using small passes and controlled welding at a set speed can produce good welds with little or no distortion. Robot-assisted and mechanical-assisted welding coupled with chill bars can help reduce and eliminate welding distortions in thin stainless steel (Figure 8.27).

Stiffened plates of stainless steel, with plug-welded stiffeners, can induce out-of-plane distortion. As the welds cool, out-of-plane distortions can be apparent due to angular rotation around the stiffener. Welding of two plates can also have angular distortions out of plane. This is the result of nonuniform stresses across the plate thickness as differential cooling occurs.

When welding stiffeners on the reverse side of a stainless steel plate, there is no good substitute for thickness. Plug welds tend to cause the opposite side of the plate to distort inward slightly due to the heated side expanding slightly more than the non-heated side. This causes a read-through of the welds to the opposite side. The higher the reflectivity, the more difficult it is to prevent its appearance.

Butt welds can be accomplished with very careful control of the heat generated into the metal. The use of chill bars or plates is paramount. Keep the stitch welds to a minimum. They tend to show as the weld pass goes over them.

Ship building defines two types of distortion in welded surfaces, in plane and out of plane. In-plane distortion is caused by shrinkage across the panel, which creates changes in the overall panel dimensions. Out-of-plane distortion is the condition where the metal surface is more wavelike and the surface rolls in and out of the flat theoretical plane. This is caused by differential expansion and contraction and by angular changes in the sheet or plate from differential shrinkage. In- and out-of-plane distortion can also lead to dimensional changes in the overall panel, because a curved and warped plate alters the overall projected flat dimensions.

[3] As discussed earlier in the book, oil canning is the colloquial term given to visible, measurable in-and-out distortions in the surface. It is not buckling per se, which is defined structurally as yielding.

FIGURE 8.27 Backup aluminum heat sink and final weld.

To overcome these distortions, require design and real-time adjustments during the welding process to reduce the development of strains near and around the weld. When designing a welded assembly, start first with the appropriate thickness. It will be far more difficult to prevent the development of plastic strains around welds – or for that matter, soldering and brazing points – on thin sheets. Depending on the fabrication, thicknesses of 2 mm minimum may make the difference needed. Move the weld to an edge if possible. It is difficult to keep distortions from appearing in welds across the plane of a surface. Angular deformation of the welded plates will cause distortions to appear.

Often small stitch welds are used to fix the assemblies together. These can help restrain the two plates as the weld is applied. They should be small and carefully applied to avoid the regular spots of deformation that appear in an otherwise acceptable seam.

Adding some curvature will aid in concealing or preventing distortions from welds from appearing. This is because curved surfaces develop more rigidity in the surface. Use of chill bars and shielding gas will aid in achieving a suitable weld. Proper fit-up is also critical. If the edges are not adequately aligned, the weld will create distortions in the surface as it creates differential stresses along the edge. Be certain that the sheets and plates being welded are leveled and free of distortion. If oil canning exists in the sheet, the welding process will accentuate it.

Heat input into sheet or plate is the major culprit in inducing distortion into a welded assembly. Keeping the heat low by utilizing chill bars, high-speed welding, laser welding, reducing the size of the weld, and skip welding techniques aid in elimination of welding distortions.

Shielded metal arc welding (SMAW) is a popular welding process for plate and heavy sheet, but it is slower and will build up heat, leading to distortion. Higher-speed processes should be considered instead, particularly those that are automated, such as submerged arc welding (SAW), electron beam welding, and laser welding. Gas tungsten arc welding (GTAW) will produce good welds in automated and semiautomated systems.

The next challenge is grinding down the weld and polishing the surface to blend in with adjacent areas. Overgrinding can develop depressions in the surface, which, due to the reflective nature of stainless steel, will show more readily. It is critical to carefully step down the grinding process to avoid going into the plane of the base metal.

Mirror surfaces are the most difficult because of the level of polish and blending that must occur. Mirror surfaces will show small distortions as you move around and view the piece. Angel Hair and glass bead are diffuse textures, which aid in concealing small distortions that may be apparent. Directional satin finishes, such as a No.4, will also help conceal distortions.

FIGURE 8.28 Oil canning on stainless steel sheet that was too thin. Some of the distortion eventually flattened as temperatures changed.

DISTORTION FROM COLD FORMING OPERATIONS

Cold forming operations such as press brake forming, stamping, and contour roll forming can induce distortions in flat plane surfaces. Assuming you begin with a flat, leveled, and stress relieved sheet or plate, most distortions in cold formed panels are due to uneven stresses induced into the material.

In forming pan-shaped panels via press brake, operator skill has the greatest impact. The process of press brake forming, unless automated, requires the operator to bring the metal up as the ram descends into the sheet. This is not difficult with a single edge, but adding a transverse bend requires care and precision. If the panel is not consistently brought into position, a small crease or distortion will appear in the corner (Figure 8.28).

The austenitic stainless steels have good ductility, particularly when they are stretched in tension forming operations, but they will distort when forming involves compression. They tend to wrinkle. If the part or form can be produced under tension forces, the result will be better than under compressive forces.

Proper design is the key to achieving a distortion-free surface. This includes designing the process of manufacture so that unwanted distortions are not induced into the stainless steel surface.

On large flat panels, inducing an outward convexity to the panel will benefit in the appearance of flatness. On large flat surfaces, there is an inherent instability. Attempting to make the surface measurably flat consistently and repeatedly requires overcoming the physical and thermal behavior of the metal. This is not possible in most cases. The thickness of the stainless steel is the most important parameter. Adding stiffeners by means of adhesives, fusion studs, and laminating boards all are a distant second choice to thickness if flatness is the requirement.

Addressing a panel already produced and trying to make it flat after the edges have been formed and the stiffeners are applied can be a lesson in futility.

Stiffeners can lock in distortion created when the adhesive cures or when they are placed into the back side of the panel. Removing the stiffeners and resetting may help, but usually only partially. Sometimes blasting the surface with glass will induce the curvature needed, but it will also matte the texture. Blasting must be performed using a very controlled process, preferably automated. When you blast a surface use only glass bead. Sand will contaminate the surface. Blasting should be performed on the side on which you wish to induce an outward bow. When the glass hits the surface, tiny craters induce tension in the surface as they stretch it out slightly. This also work hardens the stainless steel surface (Figure 8.29).

HAIL DAMAGE AND SMALL DENT REPAIR

Small dents can be removed from stainless steel. For the dent to have occurred, the stainless steel will have gone through localized plastic deformation. There may be some thinning as the dent stretches the metal slightly. If the dent is acute and has a visible crease, the crease will not be removable. If the dents are marks in the surface such as hammer marks, these as well are not repairable.

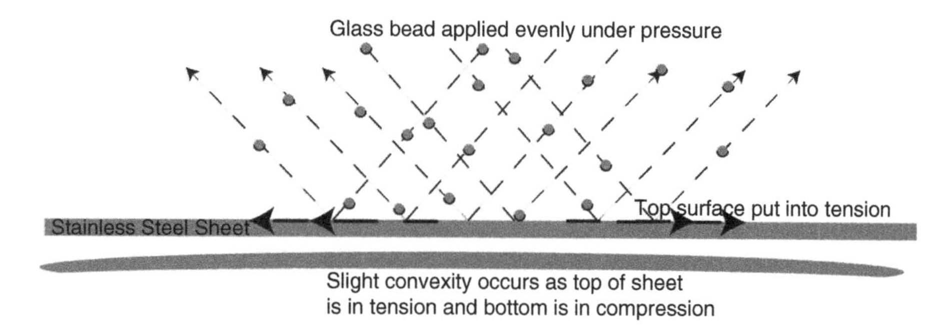

FIGURE 8.29 Glass bead blasting on one side.

To remove a dent, first lightly abrade the reverse side of the panel or sheet. This will pinpoint exactly where the dent is, and it will highlight the area.

Block the face side with a clean wood block, end grain against the metal. On the reverse side, the dent can be hammered out with a wood block against the metal. Work it gently, and the metal, if the dent is not too severe, will return close to its original position.

Table 8.7 provides a summary of the material presented in this chapter.

TABLE 8.7 Various challenges to the cleanliness of a stainless steel surface.

Impairment	Suggested remedy	Probable result
Physical Cleanliness		
Fingerprints	Isopropyl wipe Glass cleaner wipe Deionized water	Clean surface. May take more than one attempt.
Streaks and smears	Isopropyl wipe Glass cleaner wipe eionized water	Clean surface. May take more than one attempt.
Dirt and grime	Detergent Pressure wash Steam cleaning	Depends on the level and type of grime. May require hot soapy water and a nylon brush.
Adhesives	Solvents: begin with mineral spirits and move to acetone if needed.	These can often just make the adhesive gummy. The difficulty comes with removal from the surface.
Sealant stains	Ethyl acetate, methyl ethyl ketone	Similar to adhesives, these stains require loosening and rinsing. Special proprietary mixes should be considered.
Grease deposits	Detergent and steam cleaning Dry ice blasting	Should be able to dissolve the grease from the surface and rinse down.

(continued)

TABLE 8.7 *(continued)*

Impairment	Suggested remedy	Probable result
Chemical Cleanliness		
Scale	Nitric acid and potentially more intensive pickling acids	This will etch the surface. Scale should be removed at the Mill source. Castings often have thick scale that needs to be ground or blasted off with stainless steel shot.
Heat tint	Selective electrocleaning, pickling pastes if necessary	Electrocleaning operations will remove the tinting color without etching the surface. Pickling pastes will etch the surface.
Light oxidation from chloride environment	Phosphoric or chromic acid treatments, electrocleaning	This should render the surface free of the stains. Occasional rinsing will keep them from returning.
Heavy oxidation from chloride environment	Phosphoric or chromic acid treatments, electrocleaning, dry ice blast, laser ablation treatments	This should render the surface free of the stains. Occasional rinsing will keep them from returning. Some pitting will remain.
Oxidation from deicing salts	Phosphoric or chromic acid treatments, electrocleaning	This should render the surface free of the stains. Occasional rinsing will keep them from returning.
Stains from masonry cleaners	Phosphoric or chromic acid treatments, electrocleaning	This should remove the stains and leave the surface passive.
Iron deposits: transfer rust	Phosphoric or chromic acid treatments, electrocleaning, dry ice blast, or laser ablation	This should render the surface free of the stains. Occasional rinsing will keep them from returning. Some surface pitting may remain if the steel particles melted into the stainless steel surface.
Mechanical Cleanliness		
Weld splatter	Grinding, followed by phosphoric or chromic acid treatments, electrocleaning	This should leave the surface passive.
Scratches	Clean surface. Degrease, and electropolish follow with mechanical finish.	This should leave the surface passive but often shiny and may not match the surrounding finish. Blending will be challenging.
Dents: hail damage	Hammer out with wooden mallet. Suction. Dry ice followed by heat.	Small indentations may still be present.

TABLE 8.7 *(continued)*

Impairment	Suggested remedy	Probable result
Ghosting stains from packaging	If light, then try solvents to remove. If deep, then repolishing will be required.	These could be minute scratches in the surface caused by tiny abrasives in the packaging.
Welding distortion	Hammer out with wood mallet and blocking if possible. Dull finish will help conceal.	If localized, these can be reduced in appearance. If large and extending out from the weld, they are here to stay.
Cold forming distortion	Hammer out with wood mallet and blocking if possible. Dull finish will help conceal.	Similar to weld distortions, if localized one may be able to press out. Otherwise they are not repairable.

APPENDIX A

Stainless Steel Specifications Specific to Architectural and Art Applications (US and European)

STANDARD-SETTING ORGANIZATIONS (PARTIAL LIST)

American Society for Testing and Materials (ASTM)	Website: www.astm.org
European Committee for Standardization (CEN)	Website: www.cen.eu
British Standards Institution (BSI)	Website: www.bsigroup.com
Specialty Steel Industry of North America (SSINA)	Website: www.ssina.com
British Stainless Steel Association (BSS)	Website: www.bssa.org.uk
Japan Stainless Steel Association (JSSA)	Website: www.jssa.gr.jp/english
(Germany) Informationsstelle Edelstahl Rostfrei (ISER)	Website: www.edelstahl-rostfrei.de
Indian Stainless Steel Development Association (ISSDA)	Website: www.stainlessindia.org
Korea Iron and Steel Association (KOSA)	Website: www.kosa.or.kr
International Molybdenum Association (IMOA)	Website: www.imoa.info
Nickel Institute (NI)	Website: www.nickelinstitute.org
American Welding Society (AWS)	Website: www.aws.org
Steel Founders' Society of America (SFSA)	Website: www.sfsa.org

GENERAL SPECIFICATIONS

ASTM A240/A240M-18, *Standard Specification for Chromium and Chromium-Nickel Stainless Steel Plate, Sheet, and Strip for Pressure Vessels and for General Applications*

- This specification should be used when referencing stainless steel use in art and architecture. It covers the chemical and mechanical makeup of the various alloys.
  ASTM A380/A380M-17, *Standard Practice for Cleaning, Descaling, and Passivation of Stainless Steel Parts, Equipment, and Systems*
- It is unfortunate that this specification does not cover electropolishing, selective electropolishing, laser ablation, or other proven methods of treating stainless steel after welding.
  ASTM A967/A967M-17, *Standard Specification for Chemical Passivation Treatments for Stainless Steel Parts*
- This specification covers removing free iron from stainless steel surfaces. This applies to a factory or plant setting and does not cover in situ operations. There are many successful methods of in situ treatments that are not covered here.
  EN 10028-7:2016, *Flat Products Made of Steel for Pressure Purposes. Stainless Steels*
  EN 10088-1, *Stainless Steel Compositions and Physical Properties*

PLATE, SHEET, AND STRIP WROUGHT PRODUCTS

ASTM A480A480M-18a, *Standard Specification for General Requirements for Flat-Rolled Stainless and Heat-Resisting Steel Plate, Sheet, and Strip*

- This specification covers wrought alloys and finishes provided by mill producers as well as secondary finishes. It covers minimum requirements for flatness and dimensional tolerances of plate, sheet, and strip forms of stainless steel.
  ASTM A947M-16, *Standard Specification for Textured Stainless Steel Sheet* [Metric]
- This specification covers embossed and coined surfaces produced on high-pressure rolls.
  EN 10088-2, *Technical Delivery Conditions for Sheet, Plate, and Strip for Corrosion Resisting Steels for General Purposes*
- This specification describes the chemical compositions and mechanical properties of the various grades of stainless steels used to create sheet, plate, and strip.
  EN 10088-4, *Technical Delivery Conditions for Sheet, Plate, and Strip for Construction Purposes*
  EN ISO 9445-1 and EN ISO 9445-2, *Tolerances on dimensions and Form. Narrow and wide strip and cut to length for sheet and plate*

TUBING

ASTM A270/A270M-15, *Standard Specification for Seamless and Welded Austenitic and Ferritic/Austenitic Stainless Steel Sanitary Tubing*

- This is the specification to use for tube finishes. It covers only a partial list of possible finishes.
  ASTM A269/A269M-15a, *Standard Specification for Seamless and Welded Austenitic Stainless Steel Tubing for General Service*
  ASTM A554-16, *Standard Specification for Welded Stainless Steel Mechanical Tubing*
- Consult this specification when considering stainless steel structural tubing.
  ASTM A1016/A1016M-18a, *Standard Specification for General Requirements for Ferritic Alloy Steel, Austenitic Alloy Steel, and Stainless Steel Tubes*
  EN 10296 and EN 10297, *Technical Delivery Conditions for Welded Tube and Seamless Tube*
- These specifications describe the chemical compositions and mechanical properties of the various grades of stainless steels used for welded and seamless tube.

BOLTS

ASTM F593 and ASTM F594 Stainless steel specification for bolts, hex headed screws, studs, and nuts smaller than 1.5 in. in diameter.

ASTM A1082/A1082M-16, *Standard Specification for High Strength Precipitation Hardening and Duplex Stainless Steel Bolting for Special Purpose Applications*

WIRE AND ROD

ASTM A555/A555M-16, *Standard Specification for General Requirements for Stainless Steel Wire and Wire Rods*

EN 10088-3, *Technical Delivery Conditions for Semi-finished Products, Bars, Rods, Wire, Sections, and Bright Products for Corrosion Resisting Steels for General Purposes*

- These specifications describe the chemical compositions and mechanical properties of the various grades of stainless steels used to create bar, rods, and wire.
  EN 10017 Tolerances for wire rod
  EN 10218 Tolerances for stainless steel wire

CASTINGS

ASTM A351/A351M-18, *Standard Specification for Castings, Austenitic, for Pressure-Containing Parts*

ASTM A890/A890M-18a, *Standard Specification for Castings, Iron-Chromium-Nickel-Molybdenum Corrosion-Resistant, Duplex (Austenitic/Ferritic) for General Application*

BARS AND SHAPES

ASTM A276/A276M-17, *Standard Specification for Stainless Steel Bars and Shapes*

ASTM A1069/A1069M-16 Standard Specification for Laser-Fused Stainless Steel Bars, Plates, and Shapes

EN 10058, 10059, 10060, and 10061: Specifications cover tolerances for stainless steel flat bars, square bars, round bars, hexagonal bars

WELDING

AWS D1.6/D1.6M-2017, *Structural Welding Code—Stainless Steels*

- This is an important specification to use with all types of stainless steel welding.

  AWS B2 1.005, 1.006, 1.009, 1.010, 1.013, 1.014

  These are the standard procedures for welding austenitic stainless steels using GMAW, GTAW, and SMAW for stainless steel and stainless steel to carbon steel.

APPENDIX B

Relative Cost of Different Finishes of Wrought Stainless Steel

Stainless steel sheet and plate are common forms used in art and architecture. There are numerous finishes available from manufacturers around the world. Some can be combined with others. Most can receive coloring treatments to enhance their appearances.

The following graph shows the relative cost of a few of the finishes available in the marketplace. The shift in the graph shows how using interference coloring or physical vapor deposition coloring techniques increases the cost but can add value to the user.

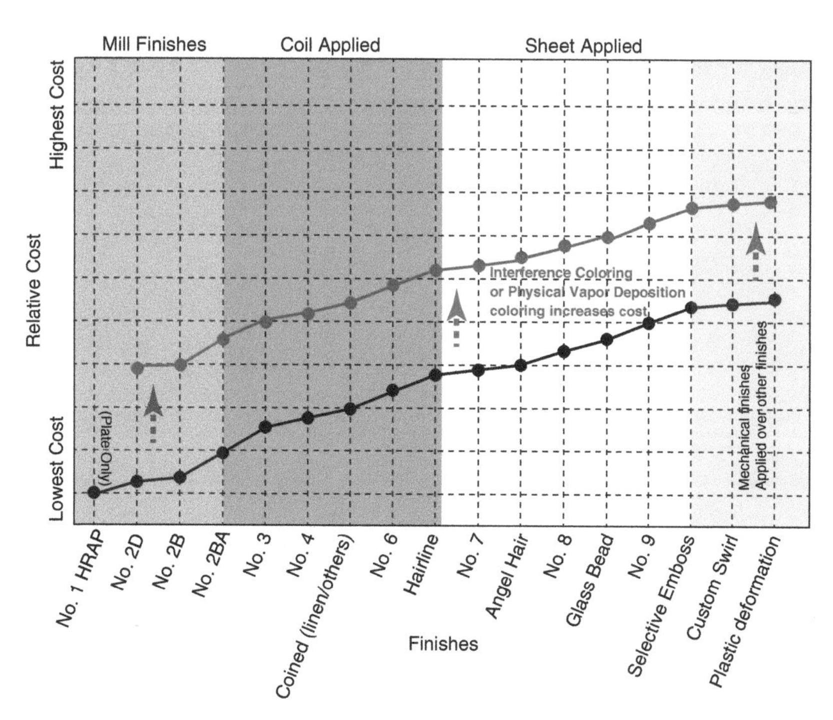

The graph of relative cost of the different finishes essentially says, for instance, that a finish such as Angel Hair is more expensive to apply than a No. 4 or a Hairline finish because these can be applied in coil forms as well as in a sheet-by-sheet form. Mirror finishes, glass bead finishes, and Angel Hair finishes are currently available in sheet or plate form. There have been some attempts to produce coil forms. The price has not been indicative of this.

Embossed finishes can be applied, at a cost, to most of the sheet finishes, but not to plate thicknesses. These finishes are not shown, but like the coined finishes they are usually added to the stainless steel when still in coil.

APPENDIX C

Relative Gloss/Reflectivity of Different Stainless Steel Finishes

It is very common for designers to consider reflectivity when using stainless steel. Some countries even limit the reflectivity to certain levels. The gloss is a component of the reflectivity of a metal surface and is determined by the way the microscopic surface scatters the reflected light or acts as a mirror and reflects the light back with nearly the same intensity as it strikes the surface.

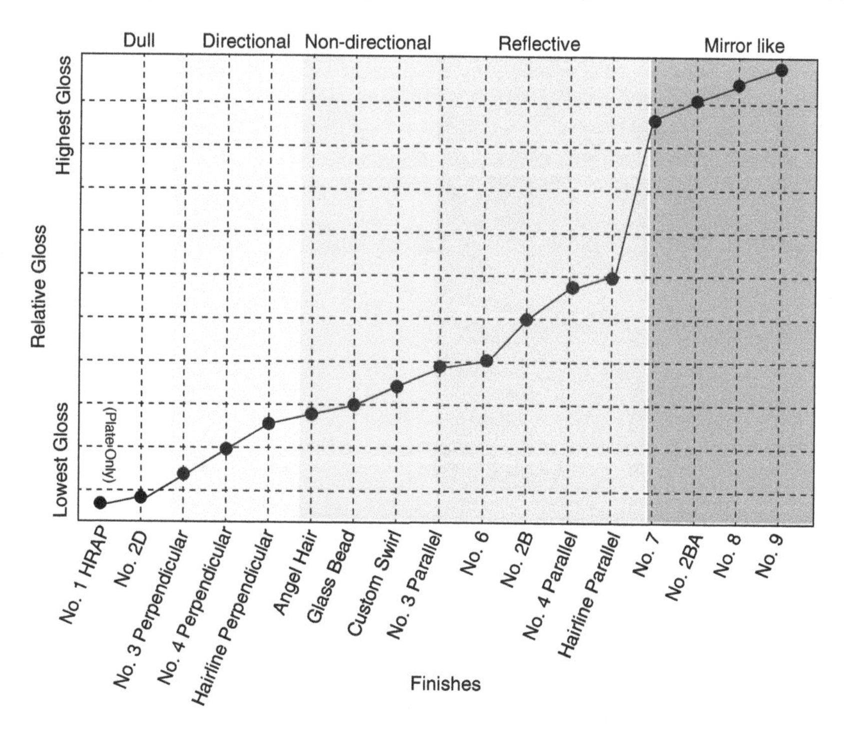

Certain finishes can be applied over others to reduce the gloss further. For example, if one applied glass bead over a No. 4 finish, the resulting gloss would place it in the dull range. Nos. 2B and 2BA are often base finishes for Angel Hair. The 2BA would provide a slightly brighter surface but still maintain a diffuse texture.

There is a distinction between the directional and non-directional stainless steel finishes. The directional finishes have different gloss readings depending on how the light strikes the surface and thus how the readings of gloss are taken. One direction will scatter the light much more effectively than the other direction. Non-directional finishes scatter the light regardless of the angle of reflectivity.

Additionally, some of the interference and physical vapor definition colors can make a gloss finish tolerable to the viewer by removing portions of the wavelength from the reflected light. There are blue stainless steel mirror surfaces and black stainless steel mirror surfaces that do not have offensive reflections.

Embossing, applied via pressure rolls can also help scatter the reflected light. It is difficult to measure the gloss reading however, because most gloss meters measure tiny regions on the surface. These regions are smaller than the embossing placed into the metal. But when a reflected surface is embossed the glare is usually reduced.

Coined finishes are similar. They can reduce the gloss but getting accurate measurements of gloss depend on where the readings are taken, the base metal or the top of the coining imprint.

APPENDIX D

List of Designers and Finishes Shown in the Book

Artist/Designer	Finish	Form	Other	Figure
Barteluce Architects & Associates	Mirror	Sheet	PVD	4.3
Bing Thom Architects	Shot blast	Cast		5.23
BNIM	Angel Hair	Sheet	Laser etched	3.42
Bowman, Caleb	Satin finish	Bar		5.13
Calder, Alexander	Shot blast	Plate		1.16
Carpenter, James	Angel Hair	Sheet	PVD	5.15
Corgan	Embossed	Sheet		4.10
De Maria, Walter	Mirror	Rod		1.13
DeAngelis, Laura	Glass bead	Sheet	Interference color	3.43
Dickens-Hoffman, Reilly	Shot blast	Cast		6.17
Elkus Manfredi Architects	Embossed	Sheet	Interference color	4.35
Elkus Manfredi Architects	Embossed	Sheet	Natural and interference	3.25
Erickson, Arthur	No. 2D, 4, 8	Sheet		3.7
Gehry, Frank	Angel Hair	Sheet		3.13
Gehry, Frank	No. 4	Sheet		3.11
Gehry, Frank	Glass bead	Sheet	Interference color	3.32
Gehry, Frank	Angel Hair	Sheet		4.2
Gehry, Frank	No. 4	Sheet		4.6
Gehry, Frank	Angel Hair	Sheet		4.17
Hadid, Zaha	Custom	Sheet		3.5
Hadid, Zaha	Angel Hair	Sheet		6.10
Handel Architects	No. 7	Sheet		3.21

Artist/Designer	Finish	Form	Other	Figure
Handel Architects	No. 7	Sheet	Perforated	5.9
Helix Architecture + Design	Embossed	Sheet	Interference color	3.34
Hendrix, Jan	Mirror	Sheet	Interference color	3.35
Hendrix, Jan	Mirror	Sheet	Interference color	3.41
Hendrix, Jan	Angel Hair	Sheet		6.3
Hilgemann, Ewerdt	Angel Hair	Sheet	Interference color	1.14
HNTB	No. 2BA	Sheet		6.1
HOK	No. 4	Sheet		7.1
Jameson, David	Custom	Sheet	Natural and interference	3.23
Kahn, Louis	Mill	Plate		3.6
Kapoor, Anish	Mirror	Plate		4.45
Kapoor, Anish	Mirror	Plate		1.12
James KM Cheng Architects	Custom	Sheet		3.26
James KM Cheng Architects	Custom	Sheet		8.8
Kohn Pedersen Fox	Angel Hair	Sheet		1.8
Kohn Pedersen Fox	Angel Hair	Sheet		3.14
Kohn Pedersen Fox	Glass bead	Sheet		3.16
Kohn Pedersen Fox	Linen	Sheet		3.27
Labja, John	Angel Hair	Sheet		6.14
Leong Leong Architecture	Angel Hair	Sheet	PVD	3.4
Leong Leong Architecture	Angel Hair	Sheet	PVD	3.36
Libeskind, Daniel	CrossFire	Sheet	Interference color	3.20
Libeskind, Daniel	No. 8	Sheet	Interference color	3.22
Libeskind, Daniel	CrossFire	Sheet	Interference color	4.8
Lippold, Richard	Mirror	Sheet	Interference color	1.15
Morphosis	Angel Hair	Sheet		4.41
Morphosis	Angel Hair	Sheet	Perforated	5.6
Moussavi, Farshid	Mirror	Sheet	Interference color	4.5
Nybeck, Beth	Custom	Sheet		3.18
Ott Architekten	Glass bead	Sheet	Interference color	3.28
Paine, Roxy	Custom	Tube		1.1
Pei Cobb Freed & Partners	Angel Hair	Sheet		3.12
Populous Architects	Embossed	Sheet	Interference color	3.30
Predock, Antoine	Mirror	Sheet	Interference color	3.31
Predock, Antoine	Mirror	Sheet	Interference color	4.4
Rojkind Architects	Angel Hair	Sheet		4.11
Saarinen, Eero	No. 3	Plate		1.5
Scuri, Vicki	Satin	Tube		6.11
Selldorf, Annabelle	Angel Hair	Sheet		6.15
Selldorf, Annabelle	Angel Hair	Sheet		4.31

Artist/Designer	Finish	Form	Other	Figure
Small, Jesse	Angel Hair	Plate		1.10
Snelson, Kenneth	Mirror	Rods		1.9
Snøhetta	Custom	Sheet		4.9
Snøhetta	Custom	Sheet		3.19
The Beck Group	Super No. 8	Sheet		3.22
Van Alen, William	2B	Sheet		1.2
Van Keppel, Dierk	Mill	Cable		5.18
Voss-Andreae, Julian	Custom	Sheet		1.11
Westlake Reed Leskosky	Angel Hair	Sheet	PVD	4.7
Young, Larry	Custom	Plate		1.3

Further Reading

Aldler Flitton, M.K. (2004). *Underground corrosion after 32 years: a study of fate and transport.* Idaho National Engineering and Environment Laboratory EMSP #86803 Annual Report.

Almarshad, A.I. and Syed, S. (2009). *Degradation of stainless steel 304 by atmospheric exposure in Saudi Arabia.* Corrosion Research Group, King Abdulaziz City for Science.

American Welding Society, Inc (1980). *Specification for Welding of Sheet Metal,* AWS D9.1-80. American Welding Society, Inc.

Avesta Sheffield (1999). *Avesta Sheffield Corrosion Handbook for Stainless Steels,* 8e. Avesta Sheffield AB.

Baddoo, N. (2013). *Steel Design Guide 27: Structural Stainless Steel.* American Institute of Steel Construction.

Baddoo, N., Burgan, R., and Ogden, R. (1997). *Architect's Guide to Stainless Steel.* The Steel Construction Institute.

Bernard, O.P. (1948). *Staybrite Steel in Architecture.* Firth Vickers Stainless Steel Ltd.

Chawla, S.L. and Gupta, R.K. (1993). *Materials Selection for Corrosion Control.* ASM International.

Cobb, H.M. (2010). *The History of Stainless Steel.* ASM International.

Coburn, S.K. (1978). *Atmospheric Factors Affecting the Corrosion of Engineering Metals.* ASTM International.

Committee of Stainless Steel Producers (1988). Cleaning and Descaling Stainless Steels. A Designers' Handbook Series, No. 9001. American Iron and Steel Institute.

Committee of Stainless Steel Producers (n.d.). Design Guidelines for the Selection and Use of Stainless Steel. Designers' Handbook Series, No. 9014. American Iron and Steel Institute.

Committee of Stainless Steel Producers (1976). Stainless Steel Fasteners: A Systematic Approach to Their Selection. American Iron and Steel Institute.

Committee of Stainless Steel Producers (1988). *Welding of Stainless Steels and Other Joining Methods.* A Designers' Handbook Series, No. 9002. American Iron and Steel Institute.

Crookes, R. (2007). *Pickling and Passivating Stainless Steel,* vol. 4. The European Stainless Steel Development Association.

Cunat, P.J. (2001). *Corrosion Resistance of Stainless Steels in Soils and in Concrete.* Study of Pipe Corrosion and Protection. The European Stainless Steel Development Association.

Dean, S.W. and Rhea, E.C. (1980). *Atmospheric Corrosion of Metals.* From a Symposium on Corrosion of Metals. ASTM International.

Delstar (2003). *Electropolishing: A User's Guide to Applications, Quality Standards and Specifications*, 9e. Delstar Metal Finishing Inc.

Fielder, N. (2013). *100 Years of Stainless Steel*. Sheffield: Self Published.

Fong-Yuan, M. and Bensalah, N. (2012). *Corrosive Effects of Chlorides on Metals*. Department of Marine Engineering, NTOU.

International Molybdenum Association (2009). *Practical Guidelines for the Fabrication of Duplex Stainless Steels*, 2e. International Molybdenum Association.

Japan Stainless Steel Association (1998). *Successful Use of Stainless Steel Building Materials*. Japan Stainless Steel Association.

Johnson, H. and Cutchin, SR. (2015). Electropolishing applications and techniques. *The Tube & Pipe Journal*.

Kalpakjian, S. and Schmid, S. (2001). *Manufacturing and Engineering Technology*, 4e. Prentice Hall.

Kerber, S. and Tverberg, J. (2000). Stainless steel surface analysis. *Advanced Materials and Processes* 158 (5): 33–36.

Kotecki, D. (2010). Some pitfalls in welding of duplex stainless steels. Sao Paulo. *Soldagem and Inspecao* 15 (4): 336–343.

Kovac, C.W. (n.d.). *High-Performance Stainless Steels*. Reference Book Series No. 11 021. Nickel Development Institute.

Lula, R.A. (1986). *Stainless Steel*. American Society for Metals.

Miettinen, E. and Taivalantti, K. (2002). *Stainless Steel in Architecture*. The Finnish Construction Steelwork Assoc.

Pauly, T., Brown, D., and Helzel, M. (2012). *Stainless Steel Flat Products for Building*, Building Series, vol. 18. The European Stainless Steel Development Association.

Roberge, P. (2006). *Corrosion Basics: An Introduction*, 2e. NACE International.

Sedriks, A.J. (1996). *Corrosion of Stainless Steels*, 2e. Wiley.

The Nickel Institute (2008). *The Nickel Advantage: Nickel in Stainless Steel*. The Nickel Institute.

The Nickel Institute (2014). *Stainless Steel in Architecture, Building and Construction*. The Nickel Institute.

Uhlig, H.H. (1948). *Corrosion Handbook*. Wiley.

Winston Revie, R. and Uhlig, H.H. (2008). *Corrosion and Corrosion Control: An Introduction to Corrosion Science and Engineering*, 4e. Wiley.

Zureick, A., Emkin, L., and Gwangseok, N. (2007). *Guideline for the Design of Stainless Steel Structures, Part 1: Dimensions and Section Properties*. Stainless Structurals, LLC.

Index

Page numbers followed by *f* and *t* refer to figures and tables, respectively.